Travel
GOALS

INSPIRING EXPERIENCES
TO TRANSFORM YOUR LIFE

INTRODUCTION

*A*t Lonely Planet we're obsessed about making travel plans. No sooner have we got our bags through the door, than we are mulling over our next adventure. That's because we are well-versed in the power of travel to change lives. We've felt that rich new perspective gained from immersion in a foreign culture, and we've experienced that deep wonder at the natural world after spending time in strange and awesome landscapes. We also know how the challenges of travel – the arriving-in-a-new-place-and-not-knowing-a-single-soul moments – teach us so much about ourselves, helping us become all the more stronger and wiser for our next trip.

In this book, we have collected together what we believe are the most extraordinary and transformative travel experiences out there. These travel goals are not about ticking off a list of far-away sights (although far-away sights do feature), they are about a life filled with variety and self-discovery. Each goal is enriching in some way, either because it's about forging stronger connections with the natural world, helping a community clear a coastline of plastic, or spending a week on silent retreat. They are not always easy.

In fact, some of them are intentionally hard; the book's structure of six chapters of increasingly more committed and rewarding goals means you are able to find a set of goals that is right for you. And have others to aim for.

So how did we go about compiling this list? We started out by asking our pool of travel writers about the travel experiences that had a life-changing effect for them – what would they recommend everyone try at least once in their lives? We got a very varied response. Some things cropped up time and again – such as solo travel and volunteering – while some were very individual, but convincing goals, such as starting a sketch journal. We then honed this list to 120 experiences that we felt would be the most personally rewarding. For each goal we explain why you should consider giving it a go and give at least three recommendations for where you can put the goal into action.

We hope this book will inspire you to think of travel as an opportunity for positive change, whether that's by learning about yourself, other cultures, or by giving back in some way. And of course to create your own set of goals that are personal to you.

NATURE
CULTURE
COMMUNITY
WELLNESS

LOCAL LIFE

SUSTAINABILITY

SELF-DISCOVERY

BECOME ABSORBED IN VILLAGE LIFE

TAKE THE SLEEPER TRAIN

TREAT YOUR BODY

KEEP A SKETCH JOURNAL

TRAVEL ON HORSEBACK

RETRACE THE STE

DINE OUT ALONE

ACCEPT THE KINDNESS OF STRANGERS

SEEK OUT SACRED PLACES

EMBRACE THE OFF SEASON

SLEEP UNDER THE STARS

PS OF HISTORY

FIND BEAUTY IN DETAIL

1

TRAVEL WIT

When you make time for the spontaneous, it opens the door to
all kinds of possibilities. Who knew you liked classical music?
Or pear ice wine? Or browsing for blue porcelain teacups in the market?

HOUT A GOAL

*J*ust about everyone has some kind of plan for their trip. You have a checklist of places to go each day: spend the morning at the art museum, climb the historic tower, then chow at the famed chef's gastropub for dinner. You use your phone's map app to tell you the quickest way from one point to the next.

But what if you tossed the plan aside? What if instead you left the day wide open and followed your nose into that bacon-wafting diner, walked into that odd museum displaying cat art, and listened to Bach in that old monastery turned concert venue. Splendid surprises pop up routinely when you make space for them. Plus, think of all the time you'll save not plotting your trip. No more trying to sort through heaps of information and agonising over online reviews. You just figure it out as you go along. Live in the moment, as the saying goes.

Spontaneous travel might feel a little scary at first, though that eases once you get in the groove of it. You can start small, maybe choose a neighbourhood to explore off the cuff, then work up to larger destinations and longer time frames.

Whatever the place, you learn about yourself venturing into the unknown. Tastes and activities you might not have considered cross your path and rouse your interest. That's when you discover you're a football fan, a blood sausage aficionado and a dead-eye shopper of antique tableware.

TAKE IT FURTHER
Be open to
human decency:
**Accept the kindness
of strangers, p44**
Engage in vigorous debate:
Haggle, p76

ROUTE DES SAVEURS
Qéebec, Canada
The Flavour Trail slices for 42 miles (68km) through Charlevoix, the pastoral region that harvests much of Québec's food. It starts an hour's drive northeast of Québec City and rolls by more than 40 farms, orchards and breweries. No plan needed, just stop wherever your taste buds command: a tomato vineyard, pear cidery, baguette bakery, sheep-milk cheese maker, or village inn that serves dishes in true farm-to-table fashion.
HOW Best for foodies, April to mid-October. www. tourisme-charlevoix.com.

ISLAND-HOPPING
Greece
Be truly impromptu and pick your island destination by whim of the ferry timetables. Head to the massive port at Piraeus, southwest of Athens, and eyeball the schedule. Some parameters help, say, a journey between two and seven hours. Then see what boat is leaving next. Ancient ruins, chalk-white villages and dramatic coastlines await. Mykonos, Santorini and their Cyclades neighbours are a particularly fine cluster for ferry-hopping.
HOW Best for adventurous types, May to August. See www.greekferries.gr.

AMSTERDAM
The Netherlands
With its glinting canals and gabled buildings, Amsterdam looks like a fairy-tale city, but that's just part of what makes it magical for spontaneous wandering. It's also compact and safe, so you can't really go astray. Pick any street (but skip the overcrowded Red Light District) and you're sure to find something intriguing: a hidden garden, a shop selling velvet ribbon or a candlelit, Rembrandt-era bar. Follow-your-nose travel rocks here.
HOW Best for urban explorers, May to August. See iamsterdam.com.

Left: Southern Aegean, Cyclades, Mykonos

Dodge peak times and you'll see a wholly different side to popular destinations – be it the beach of your wildest dreams, the cultural hotspot with Unesco World Heritage status or the city landmark that everyone is raving about right now.

TAKE IT FURTHER
Search out the unloved:
Get off the tourist trail, p64
Brave a far-flung frontier:
Cross a remote border, p156

A nyone who has travelled during peak holiday times knows the pitfalls only too well: overcrowded airports and stop-and-start motorways, sky-high flight prices and exorbitant room rates, packed beaches and rammed cities, colossal queues for museums, landmarks, theme parks – you name it. It's also no secret that some of the world's most amazing destinations are feeling the tourism squeeze. But that's no reason to stay put: being a good tourist just means getting strategic about where to go and when. Embracing the off-season is one such game plan. Beyond the obvious perks of cheaper accommodation and flights, many destinations come into their own in the low and shoulder seasons, without the nerve-fraying hordes detracting from their charms and authenticity.

If you do your research, you can apply this approach to pretty much anywhere: Paris is indeed loveable in April. The Taj Mahal, India's Unesco World Heritage magnet, sees crowds thin in September, at the tail end of the monsoon and just before the high-season rush. Italy's cultural mother-lode cities, Rome, Venice and Florence, heave and swelter in summer (especially in avoid-at-all-costs August), but they can be glorious on a crisp November day. Then there are the honeymoon favourites – the Caribbean islands, the Maldives and Mauritius, where the risk of the odd shower in the low season is a small price to pay for comparatively great deals and blissfully empty beaches. If you're prepared to pack a rain jacket and extra sweater, travel in the off season can not only relieve pressure from over-crowded destinations, but it can also show them off in delightfully refreshing ways.

EMBRACE THE OFF SEASON

GO ON A SAFARI
South Africa
It's a myth that peak-season safaris are best for wildlife-spotting. Take South Africa: low season runs from May to September, which is the country's winter. But this is actually the dry season in Kruger National Park, so though prices dip, the thinned-out bush makes for great viewing of the resident elephants, hippos, lions, rhinos, zebra and leopards.
HOW Rainbow Tours (www.rainbowtours.co.uk) offers a 16-day escorted tour, with flights. Kruger Mpumalanga International Airport (KMI) is the gateway.

TRAVEL IN THE MONSOON
Southeast Asia
Spontaneous rains lend a lush beauty to countries like Vietnam, Thailand, India, Laos, Cambodia and Myanmar during the monsoon season (roughly June to September). Temperatures are often still warm and showers come and go. The benefits of travelling at this time include cheaper rates and fewer crowds at major sites. Just be sure to pack waterproofs.
HOW You can snag excellent deals on faves such as Sri Lanka and the Maldives during monsoon times. Room rates can be half of what they are during peak season.

BEACH IT IN EUROPE
Many of us have been on a beach with barely an inch of towel space in summer, whether on Mallorca, the Amalfi Coast or the Greek Islands. Italy, for instance, heaves in August, when the whole country flocks en masse to the *spiaggia*. So go in spring or autumn for seasonal colour, ideal conditions for coastal hiking and biking, and lower rates.
HOW Winter blues? There are islands where the weather remains fine long into the darker months – and that's not just the Canaries. Try Malta, Madeira or Sicily, for instance.

EXPERIENCE THE POLAR NIGHT
It's dark, for sure, but the Arctic North is a magical place to visit during the midwinter Polar Night, when the sun never rises above the horizon. Come for wizard-wand Northern Light action, husky-sledding and spectacular icescapes.
HOW Consider a pre-winter season break to Iceland in November to beat the crowds. Aurora activity is frequent in capital Reykjavík, a great jumping-off point. Arctic Adventures (www.adventures.is) runs a four-hour Northern Lights Explorer from September to April.

SPEND TIME ON WATER

There's a special kind of feeling as you approach the shore. The scent of salt water alone is enough to soothe tense shoulders and unspool tangled minds. Then it appears in all its vastness, as breathtaking and sudden as it is expected: the undulating ocean, a tranquil sea.

On the face of such raw power and deep beauty day-to-day concerns dissipate, and in their place is an opening towards a peaceful, meditative mindset that can lead to intellectual and emotional breakthroughs. That's not to say every trip to the beach is life changing, but there is something about the water that invites escape and lures us into a state of play, something adults tend to leave behind after childhood. However, playing while all grown up improves brain function, boosts creativity, strengthens relationships and helps heal emotional wounds. And those are just the effects of playing on land, even indoors. Now imagine playing in the ocean. Surrendering to the tides, laughing and splashing with friends, parents, children. That's the secret that keep surfers, free-divers and other ocean addicts going back: it's not the sport or the adrenaline rush, it's the pure joy of it all.

STAND-UP PADDLE BOARDING

Manly, Australia

The crown jewel of Sydney's Northern Beaches has stand-up paddle board (SUP) rentals at Manly Kayak Centre. From there you can paddle pristine, sheltered coves, spot dolphins and catch stunning glimpses of Sydney Harbour. A two-hour paddle is plenty. Spring and summer (October to March) are prime time. Bring sunblock.

HOW From central Sydney, catch a ferry from Circular Quay. See www.manlykayakcentre.com.au for more info.

SWIMMING

Caneiros, Portugal

Hit the beach with the whole family in Caneiros, Algarve, where the wide, blonde beaches are backed by dramatic sandstone bluffs. The bays are calm enough for swimmers of all skills, especially kids, and day beds make the ideal base camp for the day. Avoid the summer crush; the best months are May and October.

HOW Fly into Faro Airport on the Algarve Coast, then rent a car and drive 43 miles (70km) west to reach Caneiros.

KAYAKING

Lake Tahoe, USA

A 1640ft- (500m-) deep, 22-mile- (35km-) long lake in California's Sierra Nevada mountain range is where pro snowboarders get their kayak on. When the wind is calm, the surface reflects jagged toothed mountains and visibility exceeds 66ft (20m). Kayaking is a summer sport (June to September) in the Sierras, best in the early morning or early evening to avoid sunburn.

HOW SUP Tahoe (www.supsouthlaketahoe.com), located on the southern shore, offers paddle board and kayak rentals. Fly into Reno, Nevada, a 60-mile (96km) drive away.

Left: Lake Tahoe in California

© MATT TRAIN | SHUTTERSTOCK

Activate your 'blue mind'

J Wallace Nichols, a marine biologist and author of *Blue Mind* – a book about how being in and around water can make you happier, healthier and more connected – believes we all have a 'blue mind'. He describes it as a sense of unity and a Zen satisfaction in the moment that is triggered when we're in or near water. 'We are beginning to learn that our brains are hardwired to react positively to water,' Nichols writes, 'being near it can calm and connect us, increase innovation and insight, and even heal what's broken.'

Anthropology seems to prove his point. Humans have always taken to water to relax or rejuvenate. In 2000 BC, the Ancient Egyptians practised bathing rituals to heal. Hippocrates, the father of Western medicine, advised both hot and cold bathing, and he was the first to record the use of contrast baths, which are now a staple in fitness regimes of elite athletes. Native Americans still sit in sweat lodges to purify the body and mind.

TAKE IT FURTHER

Enjoy the cool kiss of the sea:
Take a plunge in open water, p96
Take action for the future:
Become an ocean defender, p160

Towns left to ruin chime an urgent warning about humankind's effect on the planet, discovers **Anita Isalska**. But even in Chernobyl's exclusion zone, it's possible for hope to bloom.

EXPLORE AN ABANDONED PLACE

*B*roken glass crackles under my shoes as I step backwards, angling my camera towards a high-rise apartment block. Like every building in Pripyat, it's a concrete skeleton: abandoned, looted and overgrown.

Pripyat in northern Ukraine was deserted after the Chernobyl nuclear disaster. Purpose-built in 1970 to house workers at the Chernobyl power plant, the town was previously a modern, liveable place with almost 50,000 residents. But on 26 April 1986, following a failed safety test, an uncontrolled explosion tore through reactor number 4 of the power plant. After the blast, buses pulled into Pripyat to evacuate the town's residents, who were told to pack lightly and assured that they would be able to return soon. Of course, they never did.

After the explosion, radioactive material billowed into the air and was detected as far afield as Sweden and the UK. The land surrounding Chernobyl's power plant remains dangerously contaminated. Security checkpoints control access to this 'exclusion zone', a 17-mile (30km) radius of the most polluted land.

Walking into an old supermarket, I find it easy to picture the town at its prime. Looters long ago cleared the shelves, and a couple of upturned shopping trolleys indicate the need for a clean-up on aisle three. But from the design of the disintegrating chiller cabinets to the sky-blue signs above each aisle, this empty place feels familiar. Standing in an aisle, looking left and right, I sense an uncanny similarity with my supermarket back home.

Leaving the supermarket, I sidestep a clump of moss. Some of the area's highest Geiger counter readings are recorded near patches of moss, which absorb radiation like a sponge. In Pripyat, even innocuous-looking plant life is tainted.

A large Ferris wheel looms mirthlessly over the town, with rust bleeding across its spokes. Pripyat was evacuated shortly before the grand opening of the town's fairground, so this colourful amusement park has never operated. The wheel's compartments are still a jaunty shade of yellow; I imagine stepping inside and being lifted high above the beech trees and apartment blocks. Nearby, upturned bumper cars shed their paint into the weeds. It's unsettling to find desolation at a place designed for children's play.

Picking a path through the town's debris, I feel confronted by the spectre of my own impermanence.

> **"But in the bleakest circumstances, new beginnings can blossom – albeit painfully and slowly. In some apartment buildings, trees have burst through the floors."**

Suddenly and irreversibly, everyday life was ripped from the people who lived here. Witnessing the detritus of everything a community once held dear, I'm struck by the precariousness of my own existence.

Visiting a place like Pripyat, it's easy to feel despondent about humankind's capacity for harm. The immediate death toll of the Chernobyl disaster was 31 but WHO has estimated that 4000 deaths are attributable to the accident, including incidences of leukaemia and thyroid cancer that resulted from exposure to high levels of radiation. The environmental impact has been widespread and incalculable. More than three decades later, berries and mushrooms growing in the exclusion zone still aren't fit for human consumption.

But in the bleakest circumstances, new beginnings can blossom – albeit painfully and slowly. In some apartment buildings, trees have burst through the floors. Creeper plants carpet Pripyat's plazas in greenery. Many scientists have observed that animal populations in the exclusion zone are depressed since the accident, but the area isn't devoid of life: sightings of birds of prey, boar and wolves continue, offering a faint glimmer of hope in this scarred place.

Standing in this husk of a town, a place rendered uninhabitable by human carelessness, I've never felt a greater urgency to examine my own impact on the planet. Visiting Pripyat sounds the alarm regarding our ability to devastate the natural world – and I hear its warning loud and clear.

HOW Visitors to the exclusion zone can only secure the necessary visitation permits through a tour operator. Several companies offer excursions from Kyiv; Chernobyl Welcome (www.chernobylwel.com) has one- or two-day programmes (Geiger counters included). Note a trip to Chernobyl means being exposed to radiation: a similar amount as on a long-haul flight.

Take it further

Explore the difficult stories:
Learn about the darker side of history, p86
Head into the wild:
Visit empty places, p194

Top: Ruined gym with view of the ferris wheel in Pripyat, Chernobyl.
Previous page: Sands inside a building at the Kolmanskop ghost town in Namibia

WAIUTA

New Zealand

One of New Zealand's most atmospheric gold-rush ghost towns is Waiuta, abandoned to nature after the collapse of a mine in 1951. The Friends of Waiuta group resolved to snatch the site from destruction; this scattering of dishevelled timber buildings in the mist-draped hills of the South Island is now a curiosity and beauty spot.
HOW Waiuta (www.waiuta. org.nz) is a meandering 12-mile (20km) drive south of Reefton. Don't attempt to navigate the narrow, pitted road after dark.

KOLMANSKOP

Namibia

Diamonds propelled Kolmanskop to immense wealth in the early 20th century but this southern Namibian boomtown was abandoned – casino, theatre and all – when the gems thinned out and prices lost their sparkle. Uninhabited since 1956, Kolmanskop now fights a losing battle against shifting dunes: sand is piled high in the German-style houses of its heyday.
HOW Daily guided excursions depart from Lüderitz Boat Yard (Insel St); independent visitors need to buy a permit. Go early in the day to avoid the worst heat.

OLD PERITHIA

Greece

Corfu's tourism boom in the 1950s inspired numerous locals to leave mountain villages for jobs on the coast. The island's oldest village, known as 'Old Perithia', emptied and now provides a snapshot of life before mass tourism descended on Corfu. Today, it attracts curious ramblers as much for the empty heritage houses as its expansive views.
HOW Old Perithia is a half-hour drive from Kassiopi; stay overnight at lone B&B The Merchant's House (www. merchantshousecorfu.com).

If you have a long way to go, don't
unthinkingly jump on a plane.
An overnight train ride is a far more fun,
immersive and eco-friendly way to travel –
better for you and for the planet.

TAKE THE SL

Take it further

Savour and document the view:
Keep a sketch journal, p48
Cross continents and
avoid air travel:
**Make an epic overland
journey, p268**

There's something enduringly romantic about train travel. The slower pace, the gentle rhythm, the ability to surrender responsibility and just gaze out of the window. It's true of all – well, most – rail journeys. But it's especially true of the sleeper train, the hotel on wheels aboard which you can bed down in one country and wake up in another. On which you can spend long hours admiring the view, chatting to new compartment-mates, sharing bread/biscuits/tea/vodka, watching sunrises and sunsets as you roll ever on.

Of course, it's generally quicker to fly. But if you don't have to, why rush? Train travel is far less damaging to the planet: it releases the least amount of greenhouse gasses of all forms of transportation. And it's a more organic way to go. Roving by rail, covering each inch of land overground, means you're better able to understand the connections that exist between places, see the segueing of scenery as cities become suburbs, which become farmland, which become foothills or deserts or plains.

Such journeys can be done in style. There are luxe overnighting locos that channel the bygone glamour of the Orient Express or the opulence of a maharaja's palace, their wood-panelled carriages toting pianos and butlers. But you don't need such accoutrements. The greatest joy is the journey itself; even a dirty, frustratingly slow, overcrowded overnighter has its own appeal – and will certainly make memories. Squeezed into a second-class Indian bunk, unfolding your couchette as you bump across Europe or lazing back in a super-fast intercity service in China, you might not get a lot of actual sleep on your sleeper, but you'll get a dream of a ride.

EEPER TRAIN

CHICAGO–SAN FRANCISCO
USA
On its epic 2400-mile (3900km) journey, the California Zephyr tackles sheer gorges, hot desert, snowy mountains and a tonne of tunnels and switchbacks. In 1869, these tracks were the first to cross the Continental Divide, linking the Atlantic and the Pacific.
HOW The Zephyr leaves Chicago at 2pm daily. Book early for the lowest fares and best availability; tickets can be booked up almost a year ahead (www.amtrak.com).
Duration: The entire journey takes 51.5 hours non-stop.

HANOI–SAIGON
Vietnam
Nicknamed the Reunification Express when it resumed service after the Vietnam War, this 1072-mile (1726km) ride between Hanoi and Saigon runs the length of the country, providing a perfect Vietnamese diorama: historic towns, bustling suburbs, rice paddies, emerald hills and the glittering South China Sea.
HOW Choose from hard seat (cheapest) or soft, hard berth (six-bed compartment), soft berth (four-bed) or VIP cabin (two beds; book in advance). See www.vr.com.vn.
Duration: About 35 hours.

KAPIRI MPOSHI–DAR ES SALAAM
Zambia and Tanzania
A sleeper train? Or a rolling two-day safari? The Tazara Railway crosses rivers, gorges and some impressive feats of railway engineering while trundling through the enormous, wildlife-filled expanse of Selous Game Reserve, Tanzania. You might spot elephants, rhinos, lions and more from your window.
HOW Tickets can't be bought online (www.tazarasite.com); buy them at stations in Dar es Salaam or Kapiri Mposhi or via a local travel agency.
Duration: About 46 hours.

ROME–SYRACUSE
Italy
To get from the Italian capital to the historic Sicilian city of Syracuse, this overnight loco has to board a boat. Having traced the Calabrian coast, it's lifted on to a ferry to cross the Straits of Messina, before being lifted back off to continue on its way.
HOW Two sleeper trains leave Rome daily at around 9.30pm and 11pm. See www.trenitalia.com.
Duration: 12 hours, including 30 minutes at sea.

Left: Tracks through the desert near Moab, Utah

FIND BEAUTY I

Travel goals are often oriented around the big, bold bits of our world – we'll cross oceans to behold behemoth buildings, sky-scratching peaks and grand canyons – but often it's the unique and intimate intricacy of the finer things that stick in the mind's eye.

N DETAIL

© PANITHI UTAMACHANT / ALAMY STOCK PHOTO

Take it further

Cherish the journey:
Take the slow road, p132
Gain some valuable
life-lessons:
**Learn from indigenous
cultures, p120**

*Left: Masjed-E Jameh
Mosque in Esfahan, Iran*

IKEBANA

Kyoto, Japan
Sumo wrestling aside, less is more
in traditional Japanese culture. Take
ikebana (or *kado*, 'the way of flowers'),
the delicate and symbolic art of flower
arranging. See examples at Kyoto's
biannual Hanatoro festival, or attend
classes at Tokyo's Sogetsu school.
HOW Classes are from 10am to noon
Monday; www.sogetsu.or.jp/e.

PULGAS VESTIDAS

Hertfordshire, UK
Dressed fleas (aka *pulgas vestidas*) are
a traditional Mexican folk art. Often the
tiny creatures were adorned in miniature
wedding attire and presented as married
couples. See examples at the eccentric
Tring Museum in Hertfordshire, England.
HOW The Natural History Museum at
Tring (www.nhm.ac.uk/visit/tring) is free
to enter and open daily.

CATHÉDRALE NOTRE-DAME DE PARIS

Paris, France
Do you know the difference between a
gargoyle and a grotesque? (The former is
glorified gothic guttering, while the latter
is purely decorative.) Hundreds of these
handsome devils stare down from Notre-
Dame, exquisitely sculpted in limestone.
HOW The cathedral is open 8am–
6.45pm Monday to Friday; to 7.45pm
weekends. www.notredamedeparis.fr.

MASJED-E JAMEH

Esfahan, Iran
Showcasing the best that nine centuries
of artistic and religious endeavour has
achieved, a visit to Iran's largest mosque
repays time spent examining the details –
a finely carved column, delicate mosaics,
perfect brickwork.
HOW The mosque is open 9–11am &
1.15–4.30pm daily.

TRAVEL FO

Itchy feet liked to be tickled, and amid the colour, chaos and occasional calamity of any given travelling experience, there's always comedy to be mined.

R LAUGHS

*P*acking a sense of humour is every bit as essential as taking a toothbrush when you're on the road, because laughter lurks everywhere – it's a universal language that crosses borders and language barriers with eye-watering ease, and forms bonds between people of all colours and creeds. The health and well-being benefits of sharing a belly laugh are well documented (laughing decreases stress hormones and increases immunity) and travel memories are always enhanced by merriment, whether it's sought out or encountered through serendipity and surprise slapstick.

Sometimes, though, finding the funny can be the full focus of a trip, rather than simply a sideshow.

Famous festivals of wit and hilarity take place annually in perennially popular spots including Edinburgh, Melbourne and Montreal, as well as lesser-known places like Kilkenny in Ireland, Busan in South Korea and the Altitude Festival in Mayrhofen, high in the Tyrolean peaks of Austria. These carnivals of comedy all add to the attractiveness of these destinations for footloose folk seeking staged funny business, and will often influence people's travel itineraries. However, laughter can be found echoing in some of the more curious corners of the globe, too, in places where the punchlines are less predictable and the experience is altogether more immersive.

TAKE IT FURTHER

Embrace what scares you:
Feel the rush, p102
Get it wrong and
keep going:
**Have an epic
travel fail, p140**

MUSEUM OF FAILURE
Touring worldwide
There is something endlessly uplifting about other people's failures, and this touring collection of over 100 failed inventions knows it. Items range from the plain ridiculous such as laxative Pringles, Harley-Davidson cologne and pink pens designed for women to near-forgotten treasures such as the Palm-Pilot designed exclusively for tweeting.
HOW Check www. failuremuseum.com for touring dates and locations.

PATCH ADAMS
For a really immersive experience that combines travelling, volunteering and larking about, join Patch Adams (yes, the laughter-loving doctor of eponymous film fame) on a Clowning & Caring experience. Adams leads humour-based health-focused missions in Costa Rica, Ecuador, Russia, Peru, Guatemala and Morocco.
HOW Missions with Patch Adams (www.patchadams. org) last from one to two weeks. Costs cover food, lodging and in-country travel.

LAUGH THERAPY
Mumbai, India
The funny business started here in 1995, when a Dr Kataria, convinced of the therapeutic and health benefits of a good laugh, invited a handful of people to a public garden in Mumbai's Lokhandwala Complex. Now hundreds of people get the giggles at laughter clubs and yoga groups across the city.
HOW Laughter therapy sessions take place at 6.45am Monday to Saturday at Chowpatty Beach; see essenceoflaughter.com.

Left: Comedians bouncing down a hill during Edinburgh Festival Fringe

'To dine alone is preferable to any other way for me,' wrote pioneering American food writer MFK Fisher, back in the 1940s. And to think, she didn't even have a phone to stare at while she sipped her aperitif! Try it yourself for a sense of independence and sophistication as delicious as your food.

TAKE IT FURTHER

Learn to fend for yourself:
Find your supper, p58
Take your independence
to the next level:
**Take a big trip
alone, p240**

*D*ining alone has become increasingly common in the 21st century, for a number of reasons. Households are smaller. People are staying single for longer. Home cooking has become less common. But none of these explanations, true as they are, reveal anything about the pure pleasure of eating alone in a restaurant. That bit is still somewhat taboo. We are supposed to eat out in pairs or groups – candlelight dinners for two, cocktail-fuelled brunches with friends, group feasts made for Instagram. Sure, you can grab a solo sandwich during your lunch break, and sit on a park bench while you mindlessly munch. But isn't it viewed as a little sad if you dine out at a white tablecloth restaurant all alone?

Not even close. Eating alone, when done right, is all about the confidence to enjoy your own company and the hedonism of putting your entire focus on your food. The key to a successful solo dining experience is picking the right spot. Forget restaurants with menus full of trendy sharing plates or family-style dishes. Look instead to the increasing number of establishments with communal tables or bar seating. Of course, there's nothing wrong with commandeering an entire table for yourself if you feel like it. A book or a smartphone can keep you company, but why not try to be mindful, at least for one course? Sniff, savour, close your eyes. The flavours will pop like never before. Best of all, there's nobody to steal bites off your plate.

DINE OUT ALONE

BEGINNER: ICHIRAN

Japan and elsewhere
You're actually forced to eat alone at this fan-favourite ramen chain, which seats guests in one-person booths with wooden dividers, to better concentrate on the deeply porky flavour of the tonkatsu ramen. Find locations across Japan, as well as in Taipei, Hong Kong and New York City. If you're extra hungry, press the bell for a *kae-dama* (noodle 'refill').
HOW The original branch in Fukuoka is open 24/7. See www.en.ichiran.com. 5-3-2 Nakasu, Hakata-ku, Fukuoka, Japan.

INTERMEDIATE: MERCAT DE LA BOQUERIA

Barcelona, Spain
Blend in with the lunch crowds at Barcelona's iconic food market, shouldering up to stalls selling Catalan specialities such as salads with *baccalà salat* (dried salted cod) or slippery chargrilled *calçots* (a leek-like vegetable). Or come around dinner, when the market is closing and hip Barcelonéses pack in for tapas like garlicky *gambas* (shrimp) – excellent for people watching.
HOW The market (www. boqueria.barcelona) is open 8am to 8.30pm Monday to Saturday.

ADVANCED: GRAND CENTRAL OYSTER BAR,

New York City, USA
It might feel fancy, but this century-old New York institution is so full of commuters and tourists nobody blinks an eye at another solo diner. Park at the bar and slurp down a dozen icy Bluepoints. Tip: order the caviar sandwich, hidden at the bottom of the appetiser menu – you won't want to share.
HOW The restaurant is open Monday to Saturday 11.30am to 9.30pm. Reserve a table online at www.oysterbarny. com. Lower Level, Grand Terminal, 89 East 42nd St.

EXPERT: CHEZ PANISSE

California, USA
Foodies from across the globe pilgrimage to Berkeley to dine at chef Alice Waters' palace of strenuously seasonal cuisine, open since 1971. Why interrupt your culinary reverie with tablemate chatter? Lone diners are treated impeccably here, and you can always choose a more casual experience by reserving a spot for lunch instead of dinner.
HOW The restaurant (www. chezpanisse.com) is open for dinner Monday to Saturday. Bookings can be made up to a month in advance. 1517 Shattuck Ave, Berkley.

BECOME
ABSORBED IN VILLAGE LIFE

Instead of zipping around a country's must-sees, dig deep into small-town life with a long, slow stay. You'll not only experience local culture in a whole new way, you'll find yourself more relaxed than you've been in years.

*I*t's tempting to spend your vacation rushing to and fro trying to cram in maximum sights – and sometimes that's just fine. But to experience genuine cultural immersion in all its fabulous, frustrating glory, pick a town or a neighbourhood and stay put. Make local habits: take your morning cappuccino in the same cafe, shop for daikon at the same Saturday market, walk along the same river each evening, nodding to the old men feeding the pigeons. You'll get a sense for the rhythms of life in another place, and for your own internal rhythms as well.

Our churning 24/7 culture pushes constant novelty and variety, especially when it comes to travel. But behind the noise, there are immense personal rewards to travelling at a slower pace. You'll form relationships with vendors and neighbours, support the town's local economy and discover quirky, small-scale delights that don't make it into the guidebooks. The lack of pressure to cram in sights means more time for spontaneity – an impromptu chat with a taxi driver, a long wander through an unexpectedly cool market, or a spur-of-the-moment football game with local teens. Plus, staying still means a lighter environmental impact – planes, buses and trains pollute far more than bikes and sneakers.

SAOU

France

This speck of a village is smack in the middle of Provence and the Alps, and displays some of the best of both. Rent one of the ancient stone houses and spend your time wandering the mystical forest, chatting to local potters and becoming a regular at the crêperie. And, of course, buy a string bag and do your daily shopping at the local markets – don't miss the fresh lavender and goat cheese.
HOW Rent a gîte in an old marbles factory near the village centre. See www.chantebise.com.

PIJAO

Colombia

Deep in Colombia's coffee country lies the tiny, parrot-bright pueblo of Pijao. It's Colombia's only official Slow City, which means it has signed on with the Slow Movement charter, promising to promote things like farm-to-table eating, human-scale building and walkable infrastructure. Staying here means bean-to-cup lattes on the town square, long mountain hikes and sultry, silent afternoons of birdwatching followed by dips in hidden waterfalls.
HOW Meet local coffee growers and roasters on the six-hour WakeCup coffee tour (experienciacafetera.com/en/coffee-region/wakecup-coffee-tour).

NORTH ANDROS ISLAND

Bahamas

A puddle jump from the cruise-ship bustle of Nassau, Andros is the largest Bahamian island, but one of the most sparsely populated and least developed. Rent a beachfront cottage and spend your days biking between villages, exploring dozens of deserted coves, kayaking in the mangroves, meeting local woodcarvers and batik artists, and stopping for lunch at your favourite roadside stand serving conch (sea snail; a Bahamian delicacy). Just try not to come back home relaxed.
HOW Western Air (www.westernairbahamas.com) has two daily flights from Nassau to North Andros. Rolle's Place (www.vrbo.com) at Love Hill Beach has simple one-bedroom cabanas.

Left: Village life in France

© P PONOMAREVA | SHUTTERSTOCK

Travelling slow

Staying put is something the so-called Slow Movement has been promoting for years. The movement, which began in Italy in the 1980s, rejects the modern impulse to rush in favour of living life at a deliberate pace. This can mean cooking at home instead of ordering delivery, sipping tea with friends rather than texting them while you're at the gym, reading novels rather than flitting between websites. When it comes to travel, the movement encourages booking apartments or homestays instead of hotels, so you can settle in and live more like a local. Ignore the urge to see all the big sights; instead, explore nearby terrain by foot or bike. It's all about human connection and declining the kind of materialism that makes us feel like we need to do it all and have it all, right now.

TAKE IT FURTHER

Slow down and focus on a new skill:
Learn a craft, p94
Enjoy sustainable community living:
Go off grid, p192

RETRACE THE STEPS OF HISTORY

Walking helps you to slow down, focus on the landscape and notice all its lumps and bumps. Think you're just stretching your legs? Think again. You're exploring an interactive museum.

Who says you can't travel back in time? There is a way, and it involves not a jot of technological jiggery-pokery or wizardry. All you need is a pair of boots and an open mind.

Walking is as ancient as the earliest hominids, and by opting for person-power over planes, trains and automobiles, we become little different to our earliest ancestors. We're viewing the world at the same soak-it-all-in pace. Granted, in some places the surroundings may have changed, but the kernel of connection remains: put one sole in front of the other and you're shadowing soldiers, pioneers and pilgrims past.

Whenever and wherever you walk, there's history to be found. And it can enhance every outing. Almost any walk is good – the sheer act of moving can refresh mind, body and soul – but if you contemplate the backstory you'll reap even more rewards. Noting the strata of crumpled cliffs, the remains of once-mighty castles or the solemn shadows of overgrown trenches takes us back to the beginning of geological time or the horrors of past wars; it makes us question the world around us and our own tiny place in it.

So next time you set off, look closely at the furrows and the hummocks, at the seemingly out-of-place boulders and the crumbling ruins, and question who might have stood where you stand, walked the same way, centuries or even millennia before.

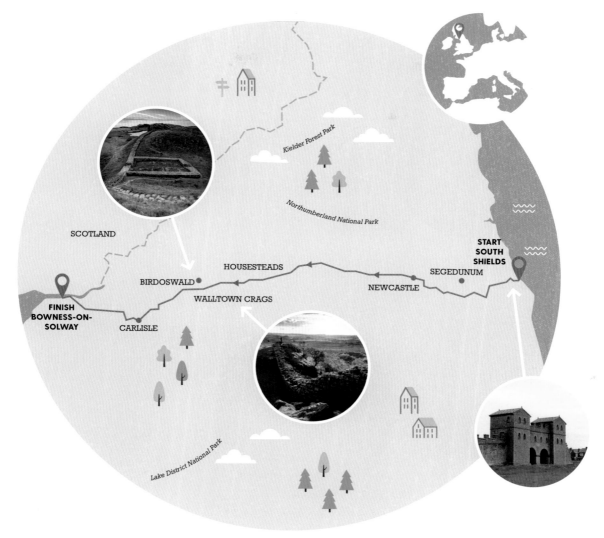

HADRIAN'S WALL PATH

UK

In AD 122, following years of skirmishing with the bothersome Picts to the north, Emperor Hadrian decided to build a beefy barrier to mark Roman Britain's furthest frontier. He picked a splendid spot: his part-stone, part-turf barricade, averaging 16.5ft (5m) high and strengthened by 16 forts, ran the breadth of the country, tracing the rolling hills and rugged escarpments of the Tyne Valley, between what is now Newcastle and the shores of the Solway Firth. Hadrian's Wall no longer exists in its entirety, but you can follow its foundations on the eponymous 84-mile (135km) National Trail, which is dotted with the legionnaire's legacy. Start at **South Shields**, the trail's eastern end, to see the excavations at **Segedunum**; admire well-preserved fortress ruins at **Housesteads** and **Birdoswald**; and tramp beside the sinuous section of stonework at **Walltown Crags**, the finest – and most atmospheric – remaining piece of Hadrian's handiwork, in a landscape that looks little changed since then.

HOW The Tyneside Metro connects Newcastle to Wallsend (the eastern trailhead). See www.nationaltrail.co.uk/hadrianswall. **Length:** 84 miles (135km). **Duration:** five–seven days.

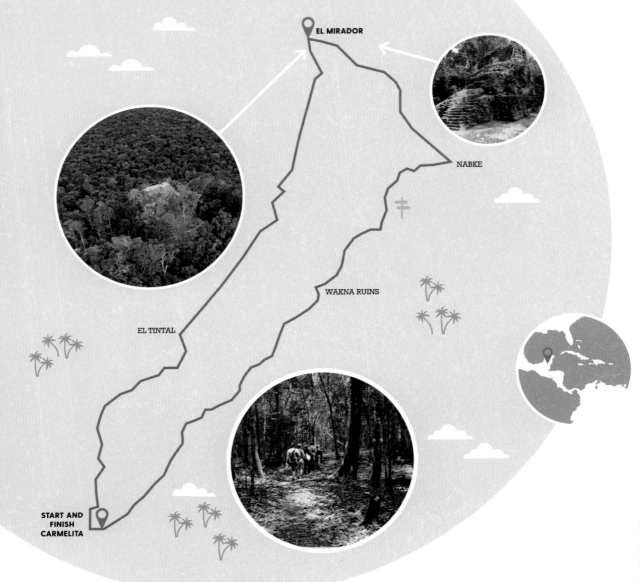

EL MIRADOR

NABKE

WAKNA RUINS

EL TINTAL

START AND
FINISH
CARMELITA

EL MIRADOR HIKE

Guatemala

The world's largest pyramids are not in the deserts of Egypt, but rather in the jungles of Mesoamerica at places such as El Mirador, a pre-Columbian complex discovered in 1926 in northern Guatemala. This Mayan centre is among the largest ever found by archaeologists and flourished from about the 6th century BC to the 1st century AD with upwards of 250,000 inhabitants. To reach it, you'll need to trek five or six days (return) along a mule trail into the dense Guatemalan jungle past the many Mayan satellite centres of the Mirador Basin. A guide is essential, while most hikers also use a mule to transport food, water and supplies. Expect to trek more than 20 miles (30km) in a day and sleep in basic campgrounds. The view atop 236ft (72m) La Danta, the largest of the three main pyramids at El Mirador, makes it worth all the sweat, as it's pure green as far as the eye can see.

HOW Arrange a trek with an authorised guide at the Comisión de Turismo Cooperativa Carmelita Agency in Flores, before heading to the starting point in Carmelita. Trekking in the rainy season, especially September to December, can be extremely difficult; February to June is the best period to attempt one.
Length: 75 miles (120km). **Duration:** five–six days.

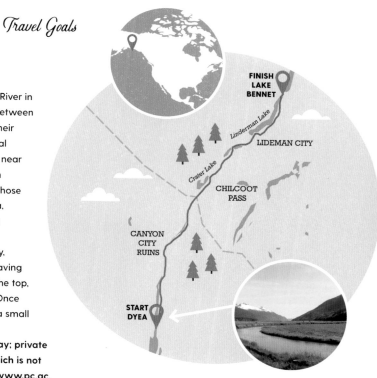

CHILKOOT TRAIL

USA & Canada

When three prospectors struck it lucky near the Klondike River in 1896, they sparked one of the greatest ever gold rushes. Between 1897 and 1898, 100,000 others followed, hoping to make their own fortunes. But it was a tough journey, not least the final overland trudge from the Alaskan coast to the gold fields near Yukon's Dawson City, crossing the USA–Canada border en route. Today, the 33 miles (53km) Chilkoot Trail follows in those hardy stampeders' boot-steps. From the trailhead at Dyea, the route wends through bear-roamed forest, over creeks and via rusting gold-rush remnants. These are especially evident as the trail hits 3525ft (1067m) Chilkoot Pass; many, faced with the intimidating climb, simply gave up here, leaving their possessions behind. It's worth the effort, though. At the top, the trail descends, finally finishing at quiet Lake Bennett. Once a teeming settlement, it is now home to scraps of metal, a small church and the ghosts of prospectors past.

HOW The Dyea trailhead is 10 miles (16km) from Skagway; private shuttles run the route. The trail ends at Lake Bennett, which is not accessible by road; access is by train or floatplane. See www.pc.gc.ca/en/lhn-nhs/yt/chilkoot.

Length: 33 miles (53km). **Duration:** three–five days.

BERLIN WALL TRAIL

Germany

The Berliner Mauerweg – or the Berlin Wall Trail – is a chilling trek back to the Cold War. It follows the infamous barricade that, from 1961 to 1989, surrounded the democratic enclave of West Berlin, splitting a city, a nation, essentially a whole continent: this ugly rampart of concrete and barbed wire came to symbolise the era's ideological divide. The wall measured around 90 miles (145km), slicing streets and squares, bisecting lakes, disconnecting neighbourhoods and turning peaceful parks into 'death zones'. It's sobering but fascinating to walk the line now. Easy-to-follow sections run via rejuvenated Potsdamer Platz, which was flattened during WWII; the notorious Checkpoint Charlie border crossing; the graffiti-covered East Side Gallery, the longest remaining chunk of wall; the Gedenkstätte Berliner Mauer memorial, which contains the last remaining fragment of wall that still has the preserved grounds behind it; and via innumerable plaques that tell stories on the spots where escapes were attempted and, often, horribly thwarted.

HOW The trail is divided into 14 sections of between about 4 miles (7km) and 13 miles (21km); the beginning and end of each section is accessible by public transport. See www.berlin.de/mauer/en/wall-trail.

Length: 90 miles (160km). **Duration:** 10–14 days.

Not only is travelling at home often more economical and eco-friendly than going abroad, it also bestows a whole new perspective and appreciation for what surrounds you.

BE A TOURIST IN Y

TAKE IT FURTHER
Give back to your own community:
Volunteer at home, p74
Explore what inspires you:
Write a travel blog, p158

*Left: Road tripping
through Death
Valley National
Park, USA*

OUR OWN COUNTRY

ou don't have to journey to far-flung lands for novel adventures. Many are on your doorstep. It's just that you overlook them because the landscape is so familiar. You forget that whales swim in your nation's waters, and that your country has historical villages and scenic trails that cross its length.

Why not take advantage of them? There are heaps of practical reasons to be a tourist in your own country. It helps the environment, since staying close to home means skipping the lengthy, carbon-emitting flight. It's generally cheaper to travel domestically, since you don't have to pay for long-haul transport or visas. Plus, there are no exchange rate costs and no pricey vaccines to get.

Mostly, though, travelling in your own country presents an opportunity to look beyond the ordinary and get a fresh viewpoint on where you live. It re-ignites your senses, teaches you new things about history and culture, and provides a sense of place that you won't get any other way.

For best results, think a bit outside the box for domestic jaunts. Try home swapping with a friend in a different city for a weekend, or letting a travel company arrange a mystery trip. Revisit a favourite holiday spot from your youth, such as a resort or amusement park. Join a local tour wherever you go. The idea is to step out of your usual role as citizen and be as inspired as a visitor.

TAKE A RETRO ROAD TRIP
USA
It doesn't get much cooler than road tripping in a 1968 cherry-red Mustang convertible or powder-blue Cadillac Eldorado along the open highway. Rent a vintage car and head out on a classic drive in the USA.
HOW Check DriveShare (www.driveshare.com) for vehicles in Chicago to embark on Route 66, or in Los Angeles to hit the Pacific Coast Hwy.

GO CAMPING ON THE SUNSHINE COAST
Australia
Many travellers have fond memories of childhood camping trips. Stoke the nostalgia with a jaunt to an old-school campground in a beachy resort area such as Australia's Sunshine Coast.
HOW Fly to Sunshine Coast Airport and drive 20 miles (32km) north to Noosa River Holiday Park (www.noosaholidayparks.com.au) for tent and RV sites.

VENTURE ON A MYSTERY TOUR
USA
Sign up for a journey without knowing the destination. Several companies specialise in mystery trips. You tell staff your budget and interests, and then they plan it. You receive info on what to pack, but the location remains a surprise until the day of departure.
HOW Pack Up + Go (www.packupgo.com) organises three-day/two-night trips for US destinations.

GO ON A DOMESTIC SAFARI
Scotland
Plenty of exotic wildlife prowls your homeland – like the Scottish wildcat in the UK – but you might not have ever seen it. Local outfitters can make it happen.
HOW Wild Highland Tours (www.wildhighlandtours.co.uk) heads out to view wildcats, red stags and more on the remote and beautiful Ardnamurchan Peninsula, a 3.5-hour drive northwest of Glasgow.

After a lengthy flight, you can sit around feeling all tired and dopey. Or you can harness jet lag's mojo and go out and explore. The sleepiness adds a wonderfully disorienting edge to your walkabout, so the landscape seems extra psychedelic and dreamy. But the enchantment only lasts until your first deep doze...

SEE A CITY IN A JET-LAGGED HAZE

BROWSE A MARKET

Markets are chock-full of sights and sounds to jump-start weary senses. Almost every city has a corker, from Marrakesh's medieval-era labyrinth of souqs to Tokyo's smelly fish market, to Istanbul's colourful Grand Bazaar.

HOW Marrakesh's souqs generally open from 10am–7pm. Visit Souqs Smata and el Kebir for a range of everything, from pottery to metal crafts and leatherware.

VISIT A NIGHTLIFE ZONE

Talk about trippy: so many flashing lights and people of the night! London's West End, with its blazing billboards and theatre marquees, and Bangkok's Sukhumvit, with its go-go bars and sizzling food stalls, are exemplars of the genre.

HOW Head to London's Piccadilly Circus, a public square that buzzes with people, commerce and neon signs at night.

CRANE YOUR NECK AT A SKYSCRAPER

Enormous feats of architecture impress even a groggy, half-awake brain. Wander around downtown and look up at the cosmos of high-altitude towers. New York City's Financial District and Shànghǎi's Pǔdōng district show how it's done.

HOW Check out the views from NYC's One World Observatory (www.oneworldobservatory.com), which spans levels 100 to 102 of the tallest building in the western hemisphere. Open 9am to 9pm, with extended hours in summer.

CLIMB ABOARD PUBLIC TRANSPORT

Grab a seat on the local boat, bus or train and watch the city drift by out the window. Sydney Harbour ferries, Lisbon's hill-conquering trams and Chicago's clattering 'L' trains all put on excellent shows.

HOW Choose a transport type and take it to the end of the line, such as Lisbon's tram E28 from Martim Moniz, which passes through Alfama.

Left: Sunrise over Shanghai, China

TAKE IT FURTHER

Break a routine and grow:
Try something new, p66
See the world from
a new perspective:
**Have adventures
with children, p68**

© 651271085 | 500px | GETTY IMAGES

EAT LIKE A LOCAL

STREET EATS

Georgetown, Malaysia
The first rule of eating local: eat where the locals eat. If the locals eat streetside, you do too. Some say that Georgetown, Penang, has the word's best street food, and hawker-style dishes like *char koay teow* and *assam laksa* are simply bowls of spicy heaven. Visit a hawker centre to do a mass sampling.
HOW Visit the locals' choice: Lorong Baru (New Lane) Hawker Stalls in Georgetown.

HOME COOKING

India
Restaurants have their place, but everyone knows homemade tastes better. When you're travelling, you make a real connection when you sit down to break bread at a local home. India's multitude of traditional cuisines – Goan, Punjabi, Tamil, Kashmiri – reach their highest expression at the tables of home cooks across the country.
HOW Authenticook (www.authenticook. com) offers meals at a home in 30 cities.

PINTXOS CRAWL

San Sebastián, Spain
San Sebastián's Basque food culture is the stuff of fine-dining legend, but some of the town's best cooking happens in the tiny pintxos bars of the Parte Vieja. Stop for the house speciality at as many of these hotspots of conviviality as you can handle, and make a meal of the mini masterpieces.
HOW Devour San Sebastián (www. devoursansebastian foodtours.com) offers pintxos bar tours in small groups.

TAKE THE PHO TRAIL

Ho Chi Minh City, Vietnam
Though pho originated in the North, it's Vietnam's unofficial national dish and beloved the country over, especially at breakfast. Join the locals for a morning bowl at one of Ho Chi Minh City's best pho joints and get an up-close view of life in an Asian metropolis.
HOW Start a walking tour with Saigon Street Eats (www. saigonstreeteats.com) with a pho breakfast.

TRY COOKING THAI

Bangkok, Thailand
Take a glimpse behind the scenes of Bangkok with a cooking class at the Klong Toey slum with local legend Khun Poo. Your class starts with a visit to the enormous wet market, provides a chance to connect with the local community and decodes classic Thai dishes like tom yum and green papaya salad.
HOW Book a class online at www.cooking withpoo.com; choose from a different menu every day.

Food is life's great leveller. When we're hungry and seated at a table, we're a single people with a common purpose. Eat with the locals and you do more than fill your belly and support a local community. You make a connection – travel's ultimate purpose.

TOUR QUEENS FOOD
USA
Step off the beaten path for unforgettable meal. Make a beeline for neighbourhoods with a history of immigrant settlement – London's East End, the 18th arrondissement in Paris; Queens in New York. These multicultural melting pots are treasure troves of culinary discovery.
HOW Discover Long Island City on a tasting and cultural walking tour with Queens Food Tours (www. queensfoodtours.com).

BOUCHONS, LYON
France
There are countries the world reveres for their food culture, and within those countries are regional hotspots that the natives revere. For the Japanese it's Osaka, for the Italians it's Bologna. For the French, it's Lyon. Heed the local wisdom, pull up a chair and discover classic Lyonnaise cuisine.
HOW Try the food of Florence Périer at Café du Peintre (www. lecafedupeintre.com) at 50 boulevard des Brotteaux.

CANTEEN CUISINE
Buenos Aires, Argentina
The hunger of working men and women has long been filled by home-style restaurants with hearty portions and low prices. In Buenos Aires' working-class La Boca it's the classic *bodegón*, serving beef grilled on the *parilla* and wine in a *pingüino* (penguin-shaped pitcher).
HOW Working man's cantina El Obrero, Agustín R Caffarena 64, hasn't changed since the 1950s.

TASTE THE SOUQ
Fez, Morocco
Dodging donkeys and haggling with vendors is part of the fun of Fez's souqs. Dried fruits, vibrant spices, fresh herbs and vegetables, more olives than you knew existed – these maze-like medieval markets are a sensory overload to stimulate the appetite.
HOW Take a tour of the market to learn how to buy the best produce, then transform it with a class at Café Clock (www.cafeclock.com).

SUNDAY SOUL FOOD
Brazil
There's something special about Sunday lunch and the tradition in Brazil calls for *feijoada*, a carnivore's delight of black beans, beef and pork. In the most authentic places it's served in clay cauldrons, with a variety of cuts such as pork ribs, smoked sausage and pig's feet.
HOW Enjoy a Sunday feast at Casa de Feijoada (restaurante casadafeijoada.com. br), Rua Prudente de Moraes 10, Ipanema.

TAKE IT FURTHER
Take a seat at their table:
Live with locals, p82
Challenge yourself to a plant-based diet:
Go meat-free on the road, p236

BE TRANSPORTED BY THE HUMAN VOICE

Chanting monks, harmonic nuns and mountain yodellers are among those who have the power to get into your ear and carry you away.

Hearing how other cultures sing is a great treat of travel. It gets pretty incredible out there. Consider the throat singers of Mongolia. Each performer creates multiple notes at once using a form of circular breathing for a sound that's both froggy and haunting. Switzerland is renowned for yodelling, where performers rapidly change their voice between low and high registers. The technique was originally used to herd cows. In South Korea, pansori singers can go on for hours in opera-like concerts using loud, dramatic voices that take a lifetime to train. Qawwali singers in India and Pakistan keep it shorter but have been known to fall into a trance-like state as they build melodies to an ecstatic release; the music is meant to bring listeners closer to God.

In general, churches, temples and other religious institutions are good places to seek out exquisite voices – which you should do, because it helps the body as well as the soul. Research shows that listening to music you connect with slows your heart rate, reduces blood pressure and decreases stress. And while music in general can do the trick, there's something about the human voice that strikes us on a deeper, more primal level.

SANTO DOMINGO DE SILOS MONASTERY

Burgos, Spain

The monks from this monastery made the British pop charts in the mid-1990s with their Gregorian chants. Listeners went wild for the monophonic, minor key incantations.
HOW Hear them each evening at the 7pm vesper service; the abbey is 37 miles (60km) south of the town of Burgos. See www.abadiadesilos.es.

FULL GOSPEL TABERNACLE CHURCH

Memphis, USA

You know legendary soul singer Al Green, the silky-smooth voice and crooner of hits like 'Let's Stay Together'? He's now the reverend at Full Gospel Tabernacle and presides over a powerful choir.
HOW Hear the vocals at his South Memphis church at 787 Hale Rd each Sunday at 11.30am. Visitors are welcome.

HAZRAT INAYAT KHAN DARGAH

Delhi, India

Make your way through the Nizamuddin slum to the shrine of Sufi musician Inayat Khan on any Friday evening, and you're rewarded with local qawwali singers sending spiralling, hypnotic melodies to the heavens.
HOW The Hope Project offers 1.5-hour walking tours of the neighbourhood that end at the shrine; music begins at dusk. See www.hopeprojectindia.org.

WESTMINSTER ABBEY

London, UK

No surprise that this soaring church – the site of royal coronations, burials and weddings – fills with soaring voices. The 22 boys and 12 men of the Westminster Choir sing psalms and canticles in thrilling harmony.
HOW Attend evensong at 5pm weekdays (except Wednesday) or 3pm weekends. The service is free and lasts 45 minutes. See www.westminster-abbey.org.

Below (left and right): Mongolian singer in traditional dress; choir boys from Westminster Abbey

© PETER LANGER | GETTY IMAGES. © PAUL GROVER | ALAMY STOCK PHOTO

Heavenly Harmony

"Tiny Stavropoleos Church sits weirdly in the middle of Bucharest's bar district. Doleful, gold-painted icons stare out around the facade, and when walking past, it was these that grabbed our attention. When we saw a couple of locals tug on the carved wooden door to enter, we decided to follow.

Inside it was dark, the only light coming from flickering candles, and the air was thick with incense. An older woman at the door welcomed us and put her finger to her lips. That's when we heard it: female voices singing in pure, angels-from-heaven harmony. The sound rose in the air and fell back down in a rush of silvery notes. Glorious, divine, otherworldly – name the cliché and this matched it. And yes, we had goosebumps.

Once our eyes adjusted to the dimness we could see it was four nuns wearing long black robes who were creating the sound. An Eastern Orthodox service was in progress. Two priests read scripture, and the nuns' harmonies filled the space between passages.

I'm not sure why it was so affecting. We couldn't understand any of the words. It's just that their stripped-bare vocals were mind-blowingly perfect and it connected me to something way beyond that room. I've been seeking the dreamy sound ever since but have yet to hear it again."
Karla Zimmerman

TAKE IT FURTHER

Get up and shake your tail feather:
Go out dancing, p62
Join the throng:
Get lost in a crowd, p148

SLEEP UNDER THE STARS

Nothing awakens the senses and the soul like a night up close to the intense, inky majesty of the night sky. Spend a night under the stars and reconnect to the universe in the most fundamental yet effortless way possible.

A night spent under the stars is a simple pleasure, but the feelings it inspires are rich and complicated. It can remind us of who we are in the most basic way, since it makes us feel vulnerable, alone before the universe, but also powerful, since it connects us deeply to the natural world.

And there are also clear biological reasons why it satisfies some of our most fundamental human impulses to disconnect. As humans we all have an internal clock called a circadian oscillator. Modern life, with its dependence on tech, muddles with this clock, and messes with melatonin levels, a hormone produced by the pineal gland and which regulates the time we switch on and off from sleep. By increasing our exposure to natural light, and light cycles, a night under the stars helps regulate this, and to listen in a more profound way to our internal clock.

Finally, as well as feeling more rested, this exposure to the natural world can enormously improve psychological well-being and lower anxiety levels. And when lying under the night sky, it's impossible not to be reminded how glorious and vast the universe is. Think of it as a way of the universe feeding your soul, for free.

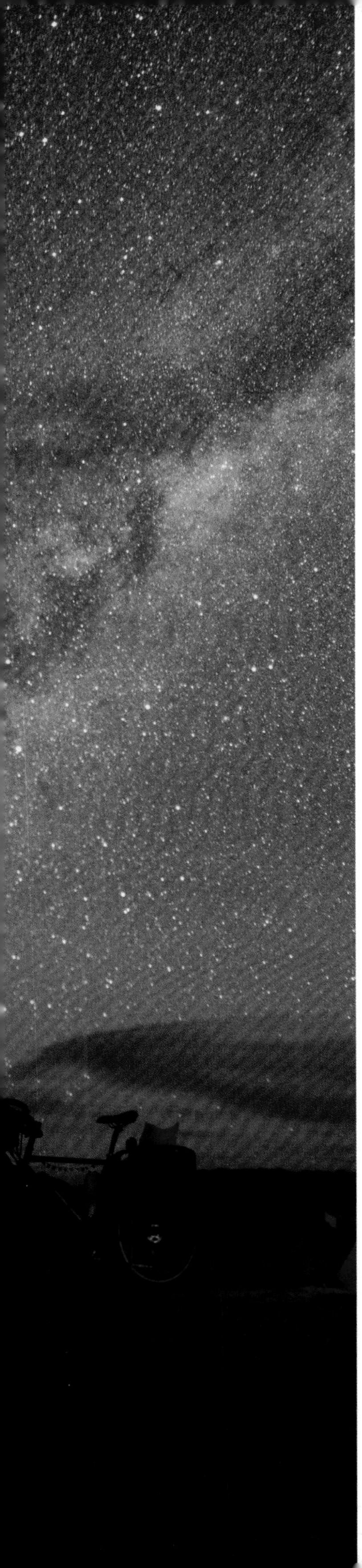

SHEPHERD HUTS

Northumberland National Park, UK
In Northumberland National Park, Hesleyside Huts are kitted out with stargazing equipment, including binoculars and telescopes, making them an ace place to admire the night skies while getting close to nature. The huts are made from reclaimed oak, insulated with sheep's wool, and have wood-burning stoves.
HOW The five huts, less than an hour from Newcastle by car, vary in size but all are charming. It's a minimum two-night stay. www.hesleysidehuts.co.uk.

DESERT CAMPING

Namib-Naukluft National Park, Namibia
Namibia is campers' heaven, and wherever you go in the country you'll find a campsite nearby. For night-sky viewing, head to the world's oldest desert, the Namib, within the Namib-Naukluft National Park. Campsites are equipped with toilets, tables and braai pits. Bring your own food and water.
HOW Park permits and camping must be prebooked through Namibia Wildlife Resorts (NWR) in Windhoek or Swakopmund. See www.canopyandstars.co.uk.

CAMPING IN THE HIMALAYA

Uttarakhand, India
Walk the foothills of the Himalaya with the *anwals*, the last migrating shepherds, moving a flock of 1200 sheep in a way of life unchanged for centuries. Spend a night under the stars at Jaikuni meadow, with mountain views and stunning night skies. Spring is the best time for this trip, though you can also do it in autumn, in reverse, when the *anwals* return.
HOW Village Ways (www.villageways.com) offers an 11-night itinerary.

Left: Camping in the Namib desert

© NOVARC IMAGES | ALAMY STOCK PHOTO

Happiness

"We cannot make happiness. We can only create the conditions by which, unbidden but welcome, our hearts swell, rise and stay buoyant because of precious, fleeting moments of joy. Lying on a blanket under the night sky to appreciate our soul's minuscule size in the face of infinity, is one way to create that condition for joy. This is because our soul is fed by experiences that are paradoxically always changing while staying the same, such as waves breaking on a beach, or the night stars that leave every morning but will always return at night. Whether we know it consciously or not, this reflects back to us an essential truth about the self and for that reason, it nourishes our soul."

Annie Pesskin, psychoanalytic psychotherapist and blogger at www.psychoanalysisinotherstories.com

TAKE IT FURTHER

Observe the unencumbered night sky:
Visit dark places, p144
Encounter the eerie:
Feel out of this world, p162

ACCEPT THE KINDNESS OF STRANGERS

Embrace it when locals invite you into their home for dinner
or offer you shelter from the elements. Then reciprocate in turn.

*A*mazing acts of kindness can often happen when we travel and, really, we're duty-bound to accept them, as they set off a chain reaction. Research has shown that kindness is contagious – that if someone is kind to us, it inspires us to be kind too. We also forge an emotional bond with people who show us generosity, and experiencing their goodness kindles our faith in others.

But for this to happen, you need to trust a stranger – which can feel awkward at first. Just keep in mind that most people want to help, and your instinct will tell you who they are. It also helps to know the local culture. Offering hospitality is part of the national etiquette in many places, such as in Middle Eastern countries where you'll invariably be invited to tea. Accept the offer and some of your most memorable travel experiences await.

FRIENDLINESS IN PORTUGAL

According to InterNations, the largest information site for expats, Portugal is the friendliest nation in the world to outsiders. Make the effort to engage people, and you'll find an incredibly hospitable and warm-hearted country.

HOW A well-placed '*muito obrigado*' (thank you), will **earn respect and smiles.**

GENEROSITY IN IRELAND

Irish helpfulness and friendliness are legendary. To get right to the heart of it, visit the local pub. You can always say thanks to new friends by buying a round of drinks. Remember: the next round should always be bought before the current round is drunk. **HOW** Practise your skills at **John Benny's** (johnbennyspub.com), a traditional pub in Dingle with terrific live music.

KINDNESS IN NEW ZEALAND

In Māori culture, *manaakitanga* (kindness) is an important value and it influences the way visitors are treated throughout New Zealand. Don't be surprised if you're offered traditional foods such as *kina* (sea urchin), *paua* (abalone; a type of sea snail) and *kumara* (sweet potato) as a welcome. **HOW** Tāmaki Hikoi's Māori guides lead cultural tours around the Auckland area. See www.tamakihikoi.co.nz.

HOSPITALITY IN KENYA

Kenyans revere visitors, and in many communities a guest will get the best provisions their host can provide, even if it's at their own expense. When eating in a local home, leave a small amount on your plate to show your hosts you've been satisfied. **HOW** Ewangan offers a homestay in a traditional Maasai village on the cusp of the Masai Mara National Reserve. See maasaimaravillage.com.

Below (left and right): Tea offered in a kulhar; Vietnamese street food

© PRAVEEN P N | GETTY IMAGES. © HADYNYAH | GETTY IMAGES

Noodles & whiskey in Northern Vietnam

"After we had squeezed out the back window of the bus and jumped to the road below, we all stared at each other. Our bus had just swerved into a mountainside in north Vietnam. No one was hurt, but the vehicle was damaged and it would take several hours for a new one to arrive. In the interim we were waiting roadside, in the middle of rice-paddy-terraced nowhere.

A boy on a water buffalo rode by. Soon he returned with an old woman with black-and-red, betel-nut-stained teeth. She motioned that we follow her around the bend to a cluster of eight tidy wood houses.

The families who lived there invited us in. Each home comprised one room with a bed and table, and no indoor plumbing. One of the families brought over bowls of noodles for everyone, while another family shared their fiery home-brewed whisky. They didn't speak English, and we didn't speak Vietnamese, but that didn't hamper an evening that included language lessons, singing and lots of laughter. They kept us warm and fed until the bus showed up about seven hours later.

Over the next few months we travelled to Angkor's jungle temples, Luang Prabang's gilded wats and other mega sights. But it's the little houses full of strangers-turned-friends that branded my memory the deepest."

Karla Zimmerman

TAKE IT FURTHER
Forge connections with someone new:
Travel with strangers, p108
Build generosity into your daily life:
Pay it forward, p126

From synagogues to streams, temples to trees, spiritual sites come in many forms and denominations. Whatever your own beliefs or faith, a respectful visit to any of them can offer real insight into what makes humans tick.

SEEK OUT SACRED PLACES

Take it further
Embark on a spiritual quest:
Make a pilgrimage, p60
Get a first-hand education about our vulnerable planet:
Learn about fragile places, p244

Y ou don't need to hold particular beliefs to visit spiritual places – or to be moved by them. The great monasteries, mosques, menhirs, mountains, rocks and rivers that humankind has come to consider sacred are, more often than not, visually spectacular. But they also have something else. An X-factor. Something that elevates them from simple soil or brick. They have the weight of history, the depth of faith, and perhaps a touch of tranquillity, a sense of connectedness or a dash of magic. Often it's almost as if you can hear the prayers of past devotees still hanging in the air.

This gives these sites extra appeal. Whether or not you believe Buddha's hair lies beneath that pagoda, or warring gods really created that lake, or prostrating around Mt Kailash in China will absolve you of your sins, there's something inherently inspiring about witnessing the faith of others. Indeed, their conviction can alter your own experience. Knowing the mythologies associated with a particular cave, cascade, tabernacle or temple makes them more than merely aesthetically appealing; they become places that can deepen your understanding of human nature.

So don't just tick off a sacred site. Feel it. Whatever your personal faith, take time to appreciate the convictions of others. Be respectful. Be accepting. Be open. You'll likely find something – be it knowledge, kinship, serenity or a moment of inner peace.

SGANG GWAAY LLANAGAAY (NINSTINTS)

British Columbia, Canada
This First Nations Haida village, on the remote Haida Gwaii archipelago, was abandoned in the 1880s. Today, it's being reclaimed by rampant forest, symbolising the Haida's connection with nature. Get there by boat, then join a Haida Watchman for tours of the sinking totems.
HOW Haida Gwaii is a two-hour flight from Vancouver. See www.gohaidagwaii.ca.

LA MEZQUITA,

Córdoba, Spain
Both building styles and beliefs clash at this magnificent mosque-turned-cathedral. Since the Spanish Reconquest, Christian iconography has been squeezed around La Mezquita's Islamic arches and orange gardens; the minaret encased in a bell tower. Now it's a Catholic basilica and a cultural icon.
HOW Entrance tickets can only be purchased from the ticket kiosk in Plaza de los Naranjos. Entry is free from 8.30am to 9.30am Monday to Saturday. See www.mezquita-catedraldecordoba.es.

CAPE REINGA

North Island, New Zealand
For the Māori, Cape Reinga – at the tip of Northland, where the Tasman Sea and Pacific Ocean crash – is the 'leaping place of spirits', where the dead descend into the afterlife. It's a place of ancestral connection, which feels as wild as the end of the world.
HOW Cape Reinga is a 264-mile (425km) drive from Auckland. The nearest town is Kaitaia, where you will find a hostel and a handful of guesthouses. See www.doc.govt.nz.

ADAM'S PEAK

Sri Lanka
Atop this pointy 7359ft (2243m) peak, deep in the tea country, is a rock imprinted with the *sri pada* – the sacred footprint. It's revered by Buddhists, Muslims, Hindus and Christians, who all make pilgrimages up the mountain and believe the mark was made by one of their own.
HOW The main trail is the 1.5-mile (3km) Hatton Path. Dalhousie is the nearest town to the trailhead; it is a 90-minute bus ride from Hatton, which can be reached by train or bus from Colombo and Kandy. See www.sripada.org.

Left: Cordoba's Mezquita, Spain

KEEP A SKETCH JOURNAL

Reaching for your phone to instantly snap something interesting or beautiful has become a habitual, even mundane, part of modern life. Instead, discover the pleasure of your own unique creativity by keeping a sketching journal of your travels.

Spontaneously sharing photographs and video clips of what we like and what we've seen is part and parcel of the digital age – it's an activity that has become central to our travels both at home and abroad. However, it has also left many of us drowning in a tsunami of images delivered daily via scores of social media apps – not to mention struggling with our own ever-ballooning online photographic libraries.

Victorian-era art critic John Ruskin was way ahead of his time when he urged travellers to lay down that newfangled contraption known as a camera in favour of pencils, paint and a sketchpad. Pay no mind to the

quality of your art, said Ruskin: drawing really is the way to 'see' a place better.

Barcelona-born Gabriel Campanario took Ruskin's words to heart when he settled in Seattle in 2006 and turned to his sketchbook as a way of making sense of his new surroundings. Two years later, Campanario founded the non-profit movement Urban Sketchers, dedicated to fostering a global community of artists who practise on-location drawing and subscribe to the motto of 'showing the world, one drawing at a time'. That simple idea has since been embraced by tens of thousands of devotees around the world who meet up to support each other in the practice of sketching.

SEATTLE
USA

The work of Urban Sketchers' (USk) founder Gabriel Campanario regularly features in *The Seattle Times*. The city supports an active USk group that meets several times a month, usually on Sunday and Friday.
HOW Venues range from the Seattle Waterfront to Swansons Nursery. See http://seattle.urbansketchers.org for the schedule.

LONDON
UK

Phil Dean leads sketching workshops in which he teaches how to capture the architecture in London with marker pen, fineliner pen and watercolour wash. He calls himself the Shoreditch Sketcher, referring to the East London area where he lives and works.
HOW See www.theshoreditchsketcher.com for upcoming events.

KUALA LUMPUR
Malaysia

With its eclectic mash-up of colonial and contemporary architecture, plus lush tropical foliage, Malaysia's capital is a captivating city to sketch. Czip Lee, the city's premier art supply shop, offers regular courses that cover the basics of outdoor sketching and how to keep a travel sketch journal.
HOW See www.artmakr.com/classes for the schedule of artist-led classes.

FLORENCE
Italy

A historic monastery in the heart of Italy's most artistic city is the base for the watercolour painting courses run by British landscape artist John Skelcher. At various different locations around Florence you'll learn about colour theory and perspective.
HOW Multiday courses are held from April to June. See www. lemarcheretreat.com for details.

Below (left and right): Drawings from Simon Richmond's sketching journals

© SIMON RICHMOND

Rediscovering a love of sketching

"In 2011 I came across a book of sketches published by Urban Sketchers Singapore. The images were so evocative and inspiring that I decided to join the group for one of its regular monthly sketch walks. In the process I rediscovered a skill I'd neglected for over 30 years since high school. Ever since, a sketching kit – black liner pens, pencils, pocket sketchpad and watercolour set – is always in my backpack, accompanying me on my travels and assignments for Lonely Planet. I also started an Urban Sketching group in my home base of Folkestone, UK.

It's now a habit for me to take time out of my research day – usually no more than an hour or so – to sketch. The practice has helped me really notice detail and 'see' destinations in a more vivid light. It's also a relaxing, fun and fantastic way to connect with locals and dig deeper into the places I write about. I've discovered people naturally gravitate towards sketchers. They pause to look at what you're doing, which can lead to a conversation and a possible new friendship. I can give individual sketches as gifts and keep the sketchbooks for myself as wonderful souvenirs."

Simon Richmond

TAKE IT FURTHER
Travel for the glory of a vibrant palette:
Be uplifted by colour, p72
Dedicate time to your imagination:
Unleash your creativity, p130

TREAT
YOUR BODY

Do as the locals do when it comes to self-care, whether that's Moroccan oil rub-downs or Chinese foot reflexology. You'll relax, feel great and connect with a different culture at an intimate level.

Visiting a spa while travelling might seem indulgent at best, a waste of time at worst. Why lie around getting a massage when you could be seeing the sights? But visiting a local spa can be one of the best ways to dig into a country's culture – while also giving your body a treat. It just depends on the spa. Avoid generic hotel facilities in favour of something distinctive to the location. In many places that means the local bathhouse. In Budapest, it's the sulphur-scented waters of the thermal springs, some dating back to medieval times. In Iceland, practically every village has a geothermal pool that serves as a relaxation and social space deep into the sunless winter. Morocco has its *hammams*, gender-segregated bathhouses where attendants will scrub you to within an inch of your life with exfoliating gloves and special black olive oil soap. Self-care could also mean a strenuous Chinese-style massage in Hong Kong, a French facial, a yoga class in India, or a Finnish sauna followed by a vigorous beat-down with a birch tree branch. Wherever you go, you'll get a sense of what it's like to be a local, which will be far more than skin-deep.

DRAGON HILL SPA

Seoul, South Korea

Koreans love a *jjimjilbang* (sauna spa), and this 24/7 multistorey facility in Seoul has everything you could ask for. Soak in the gender-segregated tubs then go for a skin-peeling body scrub. Sweat it out in the kiln saunas, including one filled with pink Himalayan salt. Then snack, drink beer and play arcade games until dawn. Ladies, you can also have your vagina steamed with mugwort if you desire.

HOW www.dragonhillspa.com/en.

WAT SOK PA LUANG

Vientiane, Laos

On the outskirts of Laos's laid-back capital is this small herbal sauna on the grounds of a monastery. Rent a sarong and step in, the air thick with fragrant steam from the wood fire. When you can take no more, head to the patio for a tea and a bendy Laotian-style massage, performed in public by chatty masseuses.

HOW The facility is about 2.5 miles (4km) south of central Vientiane – get there by bike.

FRIEDRICHSBAD

Baden-Baden, Germany

Visitors have taken to the waters of the Black Forest spa town of Baden-Baden since Roman times, and the place still has an aura of Old World opulence. Palatial Friedrichsbad is one of the most gorgeous spas in town, all marble, mosaic and soaring domes.

HOW Co-ed nudity is common. www.carasana.de/en/friedrichsbad.

Japanese sentō & onsen

Public bathing is a millennium-old tradition in Japan. There are thousands of facilities throughout the country that include *sentō* (neighbourhood bathhouses) and *onsen* (hot spring spas). *Sentō* are generally modest neighbourhood affairs, where locals gather after work to soak and chat. *Onsen*, found mostly in the mountains, range from modest outdoor pools to spectacular luxury resorts with full massage menus and delicate, delectable food. Gender-segregation and full nudity is the rule; use your tiny towel to tie around your head. Cover any tattoos carefully or you may be refused entry – ink is associated with criminal activity. And always, always take a full-shower scrub-down before your soak.

TAKE IT FURTHER

Try forest bathing:
Find peace among trees, p86
Find liberation in nudity:
Strip off, p248

Left: *Enjoying the hot mineral water of an onsen (hot spring) in Hokkaido, Japan*

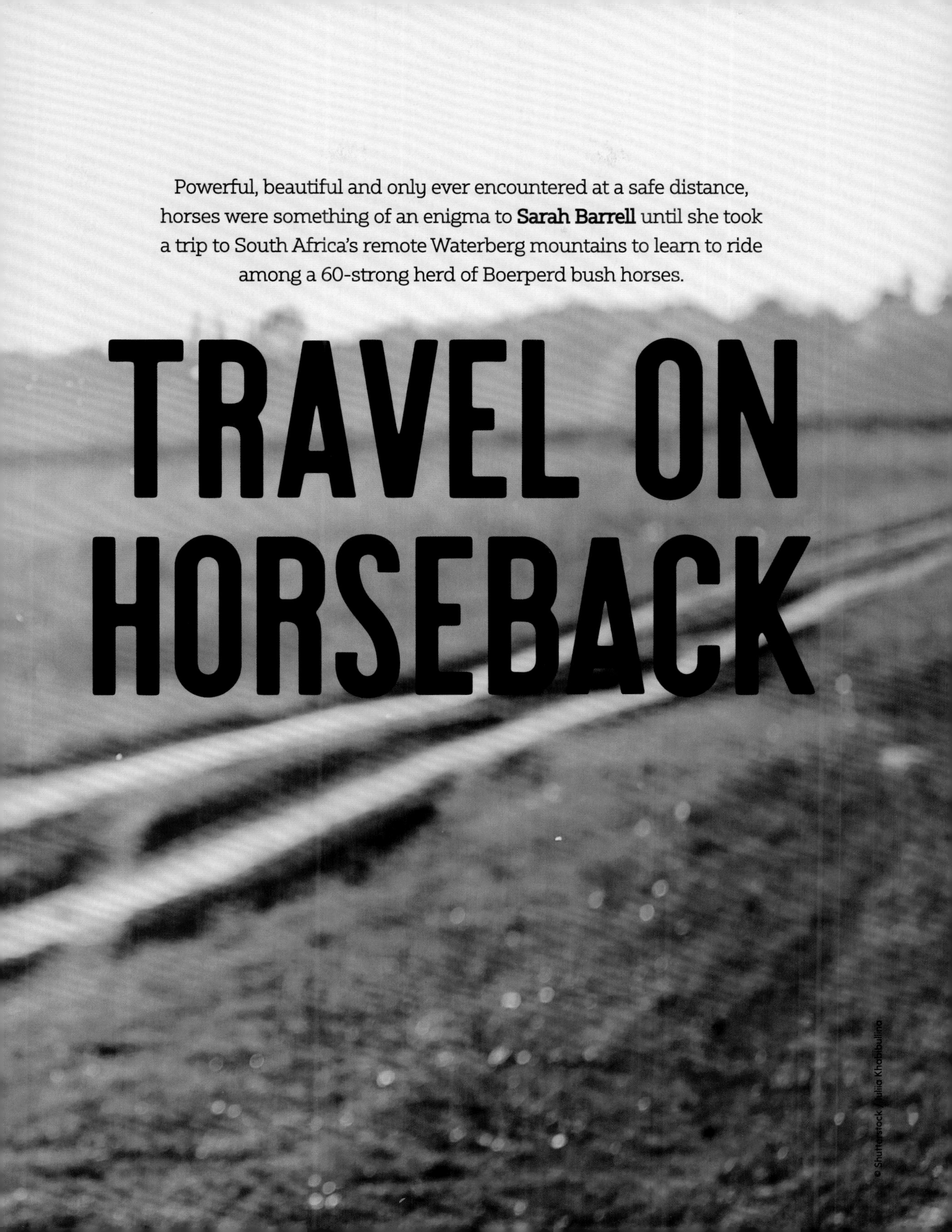

Powerful, beautiful and only ever encountered at a safe distance, horses were something of an enigma to **Sarah Barrell** until she took a trip to South Africa's remote Waterberg mountains to learn to ride among a 60-strong herd of Boerperd bush horses.

TRAVEL ON HORSEBACK

© Shutterstock / Iuliia Khabibullina

*D*iana knows what she's doing. I'm not so sure. In fact, as the rocky ground flashes faster beneath me, I'm not sure at all. I lean forwards in the saddle, an accommodating Western style that sits riders deep and gives your knees something to brace against. I relinquish any pretence that the reins I'm holding are anything other than a psychological prop, something to help me believe I'm driving this charge. I grab Diana's mane. She gives a snort, a fleeting toss of the head, and gallops onwards across the bushveld savannah. Gallops. I risk a momentary grin. I'm galloping.

Prior to arriving in South Africa's remote Waterberg mountains, my sole experience riding horses had been the sort of nose-to-tail pony convoy that passes for the sport on popular European beaches. I had what I'd termed a 'healthy respect' for these towering creatures shod in the equivalent of supersize knuckle-dusters, one that demanded I kept a safe distance. And yet the more I encountered these mercurial, conker-glossy animals, the more I wanted to understand them. And where better to do this than at the home of a horse psychologist.

Triple B Ranch, a working cattle farm set 4920ft (1500m) up in the Waterberg mountains bordering Botswana is the hub for Horizon Riding Adventures, some 60 horses and 'horse whisperer' Shane Dowinton. Famed for horsemanship clinics, and as a refuge for problem horses, the team at Horizon rears, trains and 'breaks' horses using 'round pen' reasoning and natural horsemanship techniques. 'There are no bad horses,' says Shane, 'just badly broken.' It's a rich place for learning, attracting both expert and nervous novice riders. Horizon's free-roaming herd, the majority of which are a breed of wild Boerperd bush horse, are rounded up and brought into the stables twice daily. Their thundering, bone-shaking arrival – all wiry manes and unshod hooves – renders me speechless on my first morning in the Waterberg. But intimidation quickly moves to interaction.

I'm matched to a suitable mount – the supremely steadfast Diana, a 16-year-old alpha mare whom I nickname The Travelator. And before I can say 'woah', I'm up in the saddle and off around the estate, mustering cattle, tracking game and even doing a spot of horseback yoga. The actual staying-on-the-horse-and-riding bit almost becomes an afterthought, which

is very much the point. The distractions are spectacular. One honey-bright morning a group rides up to Ghost Koppie, an outcrop of rocks above the ranch where views of the Waterberg's rounded golden-pink peaks roll on forever. Mineral-rich lichen in iridescent green and yellow coats the steep, craggy rocks; the exact colours of the chameleon we discover lurking in an aardvark hole. The horses pick their way around with mountain-goat agility; all I have to do is sit tight and avoid needle-sharp acacia trees. And the wasps... a series of squeals dominoes down our convoy as the group negotiates a narrow pass overhung by a wasp nest. Shane hops off his horse and appears with an Aloe leaf to remedy the brief, intense burning from the stings.

But it's encounters with bigger animals that leave a more indelible mark. Home to hippo, giraffe, zebra, antelope, leopard, caracal, brown hyena, jackal and hundreds of bird species, the Waterberg is premium wildlife-spotting terrain. Led by guides who sportingly act as decoy when approaching risky rhino or hippo, we get within mere metres of Africa's big game. On horseback, these animals simply regard us as another four-legged creature. Emerging out of the forest into a sudden clearing, we come upon four spectacular giraffes. They regard us briefly then go back to browsing for tender leaves. With a squeeze of my heels I edge Diana closer, until I can map the unique markings on their muscular necks, see the steam rising off their proboscis-like tongues. And so we stand, for some time: just a herd of animals basking the morning sun.

> "Their thundering, bone-shaking arrival – all wiry manes and unshod hooves – renders me speechless"

Above (left and right): Icelandic horse in the snow; horses grazing with a backdrop of Torres del Paine National Park, Chile.
Previous page: Mounted and ready

WATERBERG
South Africa

It was to these pristine mountains that Nelson Mandela first travelled when released from prison, and you can see why. Horizon Riding Adventures' Triple B Ranch sits pretty on their forested slopes, dense with big game and about 300 bird species. There's no set itinerary, but two riding activities a day might include outrides into the savannah and mountains, polocrosse, cattle mustering, jumping or swimming bareback with your horse. **HOW** Horizon is about 2.5 hours from Johannesburg International Airport. It offers everything from classic safari rides and family ranch holidays to expert horsemanship clinics. See www.ridinginafrica.com.

TORRES DEL PAINE
National Park, Chile

A ride from estancia to estancia in this national park takes in an incredible diversity of landscapes, from wild pampas to the jagged snow-capped peaks of the Andes, from soft Chilean grassland to steep, glacial-carved river valleys ribboned with grey-blue moraine and turquoise rivers, fringed with ancient forests. October to March is the best time to travel. **HOW** The gateway city/international airport is Punta Arenas, a two- to three-hour transfer to the Torres del Paine National Park. Ride World Wide offers packages suitable for intermediate and experienced riders; seven- to 10-day tours allow you to get the best of the destination. www.rideworldwide.com.

SNÆFELLSNES
Iceland

Encounter landscapes hewn in epic sagas, on the back of unique five-gait mounts cherished by Vikings. Glide over the black lava fields, and yellow-sand beaches of Iceland's remote westerly peninsula, backed by the multicoloured mountains of Hítardalur valley on sure-footed horses that, while small, are too powerful to be called mere ponies. Add a Golden Circle trek to take in the famed geysers, hot springs and volcanoes of Iceland's south. **HOW** In The Saddle offers tours for intermediate and experienced riders. A four- to seven-night trip includes both mountain and beach trails. Transfers from Reykjavík can be arranged. See www.inthesaddle.com.

HIDE OUT TO WATCH BIRDS

LEARN ABOUT THE DARKER SIDE OF HISTORY

MAKE A PILGRIMAGE

LIVE WITH LOCALS

VOLUNTEER AT

TRY SOMETHING NEW

FIND PEACE AMONG TREES

FIND YOUR TRIBE IN A FOREIGN CITY

LEARN A CRAFT

GO OUT DANCING

GET OFF THE TOURIST TRAIL

HOME

2

FIND YOU

Foraged food is the ultimate in local, waste-free
dining. What better way is there to understand
a place than through your stomach?

TAKE IT FURTHER

Dare to try a new source of
low-impact protein:
Eat insects, p146
Learn to take care of yourself:
**Survive in the
wilderness, p224**

R SUPPER

It's the antithesis to the ready meal: no premixed gloop, no additives (except maybe some soil); 'packaging' courtesy of Mother Nature. Foraging for food, and then magicking it into your own lunch, is simply the most satisfying way to eat. Not only do you get a wholesome feed, but also a primal reconnection with the earth. Foraging helps you tune into nature and the changing seasons to gain a sense that this is how eating used to be. Plus, there's also a cave-person smugness in knowing that, if all the supermarkets disappeared tomorrow, you could fend for yourself.

Autumn is bonanza time for foragers as fruits ripen on trees, berries appear as morsels of sweet delight on bushes – mind the thorns! – and mushrooms sprout overnight from the damp ground. You need only keep your eyes open and a basket at the ready. In the seas and rivers, wild fish such as mackerel and salmon are readily caught, while the coastline can be combed for shellfish and edible seaweed. Sometimes, a professional guide will help point you in the right direction but usually asking locals for tips will be enough to get you harvesting the region's specialities for a couple of hours. Just make sure you know what you are picking and take only what you will eat. There are plenty of other two- and four-legged foragers who will be relying on that same wild bounty.

PLUCK BERRIES
Sweden
The Swedes take foraging so seriously that it's even written into their law: *allemansrätten* (every man's rights) grants permission to hike, camp and berry-pick on another's land, as long as it's done respectfully.
HOW Simply head out into the country's ample empty spaces in July and August – the pick of the crops are sweet strawberries, golden cloudberries, blueberries and lingonberries. Remember to leave enough berries for the next person.

DIVE FOR SCALLOPS
Nova Scotia
Clinking masts, bright-painted huts, a salt-whiff on the breeze – the ocean defines Nova Scotia. Its waters are a veritable fish soup and, equipped with a Recreational Scallop Licence, you can dive down into the blue and take a portion for yourself.
HOW Scallop season runs from Easter to the end of July, and limits are set at 100 a day. Register for a scallop licence at www.fishing-peche.dfo-mpo.gc.ca.

PICK WILD GARLIC
UK
Between March and May, many British woodlands start to stink. Wild garlic – or ransoms – grow rampant, their shiny green leaves and cheery white flowers cloaking many an undergrowth while infusing the air with potent anti-vampiric fumes.
HOW Hike off with a pair of scissors in hand – the leaves are best snipped carefully at the stem, and are more flavoursome before the plants begin to bloom.

FUNGHI
France
In the great forests of eastern France, autumn's seasonal delicacy is the mushroom, piles of which you will see in the town markets. If picking your own it's best to stick to a couple of varieties you can identify with certainty. Start with chanterelles, which are the colour and smell of apricots, and the bulbous cep.
HOW Mushroom season runs from mid-August to mid-September. Local pharmacies are trained to help you identify poisonous varieties.

You don't have to be religious to benefit from following a traditional spiritual journey. Take a pilgrimage walk to learn more about yourself and bond deeply with others on the way.

Humans have been making pilgrimages for thousands of years, from early Christian voyages to the Holy Land, to the Muslim Hajj, to modern-day journeys of self-discovery like hiking the Pacific Crest Trail, made immensely popular by Cheryl Strayed's 2012 memoir *Wild*. As independent as we might be, there is something deeply connecting about walking in the footsteps of so many others, of following a ritual made holy through years of faithful repetition. Perhaps the best part is the companionship of your fellow travellers, bonds made while shaking out worn hiking boots or during an early-morning meditation session.

Believers, non-believers and those who aren't quite sure are all welcome on many pilgrimage routes. Whether or not you're religious, you stand to gain something special from a pilgrimage that you won't find on an ordinary hiking trail. First, there's the intense sense of community: many pilgrimage routes have communal lodgings where walkers come together at the end of the day, and conversations here tend to turn deep, quick. A casual chat about your daily mileage turns into a heart-to-heart about life goals, losses and the existence of a higher power in the time it takes to drain your canteen. Second, there's the connection with something larger than yourself, and not just on a spiritual level. You're doing something that people have done since time immemorial, and you can feel it. Third, there's a sense of accomplishment to finishing something with a ritualised endpoint, whether that's kneeling in a cathedral or scrawling a wish on a shrine's prayer card.

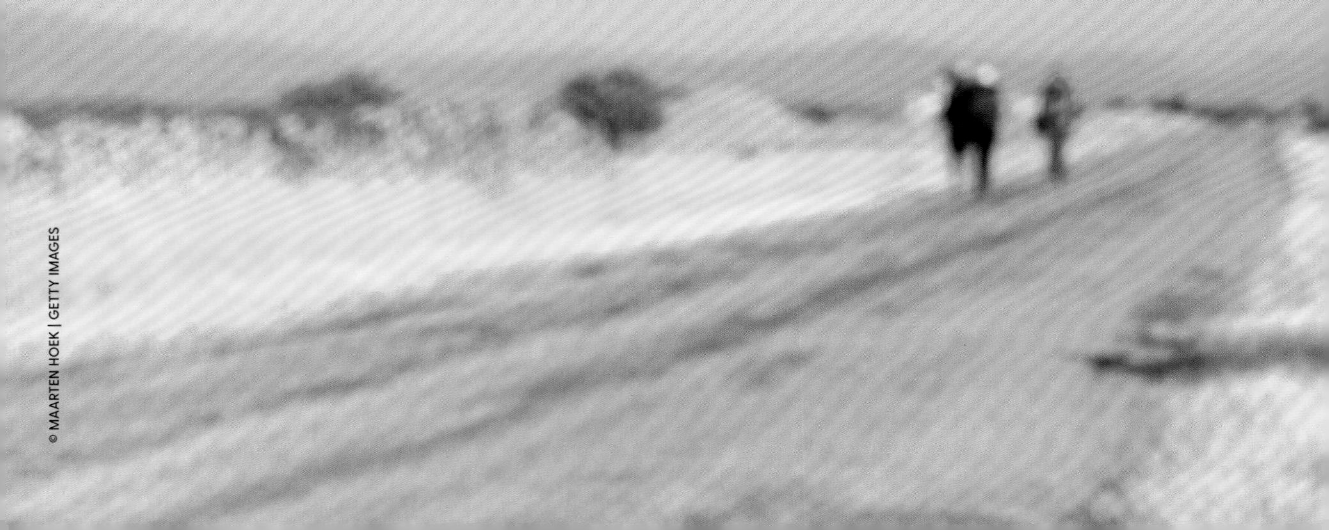

MAKE A P

THE CLASSIC: CAMINO DE SANTIAGO

Spain

Pilgrims have been travelling the Camino de Santiago (Way of St James) for a thousand years, crossing from the French Pyrenees into Northern Spain. Share spartan pilgrims' hostels at night with fellow walkers and finish at the Cathedral of Santiago de Compostela to earn a certificate, given out since medieval times.

HOW Advance bookings are useful at private hostels, but not accepted at public hostels. See www.oficinadelperegrino.com.
Duration: 35 days.
When to go: Year-round.

THE MOUNTAIN: EMEI SHAN

China

One of Buddhism's four holy mountains, Sichuan's Mt Emei is all dappled pine forest, mossy rock carvings and mist-shrouded temples. Today, mostly secular hikers brave 8000ft (2440m) of stairs, switchbacks and aggressive macaques to reach the Golden Summit. Stay in a centuries-old monastery on the way, eating breakfast with the monks.

HOW No bookings required at monasteries; no English spoken. See http://whc.unesco.org/en/list/779.
Duration: two–three days.
When to go: May–Sep.

THE LITERARY JOURNEY: PILGRIMS WAY

UK

Follow in the footsteps of the Knight, the Cook, the Wife of Bath and the rest of the pilgrims from Chaucer's *The Canterbury Tales* with a walk from London to Canterbury. The 68-mile (109km) hike follows the Thames out of the city, along busy byways and into the green countryside. End at the shrine of Thomas Becket in Canterbury Cathedral.

HOW Find pubs and B&Bs in villages en route. See www.britishpilgrimage.org
Duration: seven days.
When to go: Jun–Sep.

Below: Typical shell sign showing the way on the Camino de Santiago

GRIMAGE

TAKE IT FURTHER

Travel for what you love:
Follow your passion around the world, p116

Partake in a spiritual spring clean:
Cleanse your soul, p286

GO OUT DANCING

Nothing is quite as transporting as the sensation of moving to the music, be it line-dancing, flamenco, Texas two-step, salsa or techno at a legendary club. Music can mainline a dancer into the experience of a country or culture faster than anything else. Go ahead – get swept away.

THE CONTINENTAL CLUB
Austin, USA
The Texan live-music capital is filled with great spots in which to shake a leg, including the legendary The Continental Club and its proficient two-steppers. Head down to guitarist Redd Volkaert's cover-free Saturday show, or any day of the week when you're sure to find people whirling around in their cowboy boots.
HOW There are up to four shows daily, from mid-afternoon to 2am; check the schedule at www.continentalclub.com. 1315 South Congress Ave.

DONDE FIDEL
Cartagena, Colombia
Just off Plaza del Reloj, the blaring sounds of salsa announce Donde Fidel's incredible dancers. Squeeze into the bar's narrow spaces beneath countless photographs of music legends, but don't think you can just watch: someone will inevitably drag you to the floor.
HOW It's open from 11am to 2am daily. El Portal de los Dulces 32-09.

AMOR DE DIOS
Madrid, Spain
Flamenco is not a venture to be taken lightly, as watching a performance will show you. Get instructed by the best at Amor de Dios studio in Madrid; just make sure you're committed to going all in.
HOW Check out the Spanish-language-class timetable at www.amordedios.com. 1st floor, Calle de Santa Isabel 5.

CLÄRCHENS BALLHAUS
Berlin, Germany
Step back in time when you visit this prewar ballroom, where the couples waltzing might make you think it's 1913 again. The schedule has regular dance classes to get you up and running.
HOW Check the class timetable at www.ballhaus.de. Auguststrasse 24.

TAKE IT FURTHER
Go with the flow:
Get lost in a crowd, p148
Be a talking point:
Take a vintage transport method, p186

Many of the world's most famous tourism destinations are feeling the burden of too many visitors. Getting off the beaten track can enrich your travel experience while at the same time easing the strain on the world's over-loved places.

GET OFF THE T

It's a feeling every traveller has experienced at some point: the agony of waiting in a queue for hours to catch a glimpse of a painting or historic monument we've known intimately for years from our childhood history books, only for the view to be blocked by crowds and a forest of selfie-sticks. Plus, well, don't we all want to feel intrepid? Exploration seems to be a natural human instinct, but the rise of social media and increased access to many of the world's most famous sights have combined to put strain on certain destinations.

But that doesn't mean we should closet our suitcases just yet. Instead, the issue of overtourism is a chance for travellers to take collective responsibility for preserving the incredible heritage and history of our world, while also spreading the love by seeking out places and sights that could benefit from an increase in visitors.

Achieving this goal requires a traveller to think about each and every aspect of their trip and to be mindful about the issues facing some destinations. It doesn't mean that we have to avoid crowded cities or sights altogether, but necessitates a mix of thinking outside the box and a sense of adventure to find less-loved places. Perhaps this means visiting a destination at a time of year when the weather might at first seem prohibitive, taking a chance on an unknown neighbourhood or simply choosing somewhere totally different. By adopting this approach to our travels, we also flex our courage, open-mindedness and flexibility. And in the end, isn't that what travel is all about anyway?

OURIST TRAIL

Take it further

Enjoy the freedom of the self-propelled road:
Travel by bike, p170
Go far from anywhere (and anyone):
Disappear, p262

Left: Climbing a ridge on Mt Triglav, Slovenia

NORTHERN PERU

For an alternative to Machu Picchu, aim for Northern Peru where there's a wealth of sights with a fraction of the crowds. The pre-Incan ruined city of Kuélap holds plenty of misty mystery and the small high-altitude towns here are good for discovering Peruvian culture, cuisine and natural wonders. **HOW** Fly from Lima to Chachapoyas – the gateway town to Kuélap. Eco-friendly Gocta Lodge (www.gocta lodge.com) overlooks the Catarata de Gocta, one of the world's tallest waterfalls.

SLOVENIA

Whereas Italy's incredible sights attract huge crowds, neighbouring Slovenia can feel overlooked. Ljubljana's canals and historical architecture are a sublime stand-in for Venice, while the Slovenian coast provides the best of the Med, and Mt Triglav and Lake Bled wow just as much as the Italian Alps and Como. **HOW** International airlines and the European rail network reach capital city Ljubljana, from where a handy system of trains and buses connects travellers to the country.

TIEN SHAN, KYRGYZSTAN

In recent decades, the Mt Everest base camps (in Nepal and Tibet) have suffered from an excess of climbers (and of the waste these climbers leave behind). Similar beauty, trekking and climbing experiences can be had in the neighbouring and virtually unvisited Tien Shan range. **HOW** Kyrgyzstan's community-based tourism network (www. cbtkyrgyzstan.kg) connects travellers with local trekking guides and homestays, ensuring money remains in the local economy.

TRY SOMETHING NEW

RIDE A PENNY FARTHING

London, UK
Once simply known as a 'bicycle', the granddaddy of today's two-wheelers is a head-turner with stately Victorian style. The Penny Farthing Club holds classes where you'll learn to get on (and stay on), and take a ride through the streets of central London, perched high on your vintage velocipede.
HOW Learn more at www.penny farthingclub.com.

GO THE WHOLE HOG

Tasmania, Australia
When we kill an animal for its meat, respect demands that we don't waste a thing. Rare-breed free-range pigs, sustainably farmed in Tasmania's Derwent Valley, make the ultimate sacrifice for a 'nose-to-tail' cooking class where you'll learn to prepare every part of the pig – leaving nothing but the oink.
HOW Take a two-day whole hog course at the Agrarian Kitchen. See www.theagrarian kitchen.com.

WARRIOR TRAINING

Mongolia
More than a million nomads live in Mongolia's vast wilderness. Get an insight into their unique way of life and an immersive history lesson via a Mongol warrior training course. Learn everything you need to join Genghis Khan's army, from archery to lassoing and shooting from horseback (all in warrior costume).
HOW Book at www. nomadstours.com; tours depart from Ulaanbaatar.

FLEECE TO FELT

Scotland
Travel back to when making, not buying, was how to acquire new things. In the crofting country of the Scottish Highlands you can learn the ancient techniques of felting, from shearing a Shetland sheep to colouring with plant-based dyes and making felt artworks.
HOW Craft courses at Wild Rose Escapes (www.wildrose-escapes.co.uk) are held at Crochail Wood, not far from Loch Ness.

CLIMB HIGHER

USA
Were you a childhood daredevil, always hunting for higher branches to clamber up? It's never too late to revive your inner adventurer. Tree-climbing classes in Oregon and around the USA will give you the technique and courage to hit the high boughs.
HOW Four-day courses with Tree Climbing Planet (http://. treeclimbingplanet. com) are held at several US locations.

TAKE IT FURTHER
Push the boundaries of
your comfort zone:
**Experience culture
shock, p172**
Discover your courage:
**Face your fears,
p266**

'A change is as good as a holiday,' they say. Surprise yourself by learning something new, trying something you've never tried before, doing something crazy. You might find a new driving passion.

*Above: Harris's Hawks
are sociable birds and
popular with falconers*

LEARN FALCONRY
England
Get up close with some of nature's most majestic and awe-inspiring creatures when you take a falconry lesson in the vast, wild moorlands of Dartmoor National Park. You'll work with owls, falcons and eagles in a practice that's thought to date back thousands of years.
HOW Dartmoor Hawking (www. dartmoorhawking. co.uk) offers half-day bird of prey experiences.

URBAN FORAGING
San Francisco, USA
City-dwellers tend to do most of their scavenging in the supermarket, overlooking the wild bounty that's on their doorstep. Grasslands, creek banks and urban parklands are potential sources of wild greens, mushrooms, edible flowers and any number of foraged delights. You just have to learn how to look.
HOW In San Francisco, ForageSF (www. foragesf.com) offers mushroom and wild food classes and walks.

SKATE THE SAND
Dubai, UAE
No snow for your board? No problem – it's a transferable skill. Visit the massive sand dunes outside Dubai where you can enjoy the adrenaline rush of boarding while wearing a T-shirt and jeans. The Red Dune, which is about 30 miles (50km) from Dubai, is the Chamonix of sandboarding.
HOW You'll find sandboard rentals near the Red Dune. Go at sunrise or sunset to avoid the daytime heat.

TRUFFLE HUNTING
Italy
They're the gold nuggets of the culinary world, precious hidden morsels that come with a high difficulty rating and a price tag to match. Hunt one down with a local guide and his truffle-sniffing dog in Italy's Piedmont region, the epicentre of Slow Food.
HOW Reserve your place at Le Baccanti (www.lebaccanti. com). The hunt in the woods near Alba takes place each October and November.

JOIN A DIG
Various
Fascinated by the past? Travel to dig deeper. Archaeological digs in countries around the world – from Italy to South Africa to the Americas – need volunteers to join their search for buried treasure. Be part of a scientific community, learn new skills and get your hands dirty with history.
HOW Browse archaeological volunteer opportunities at www.pasthorizons. com/worldprojects.

HAVE ADVENTURES WITH CHILDREN

Step outside your comfort zone with children and adventure travel will reveal a world of challenges and rewards to create a lifetime of memories. According to **Clover Stroud**, it might just be easier than you think.

*A*fter the miraculous joy of a new-born baby, 'normal life' with kids can sometimes feel like a relentlessly hard grind. I'm lucky to have five children, with 16 years between my eldest and youngest. While I have the briefest memory of the thousands of school runs I've done, the scores of sports days, nativity plays and school concerts I've dutifully sat through, the memories I can recall most vividly are the times I've taken my kids out of our usual routine and gone on an adventure.

Adventures with children can be anywhere and it doesn't have to cost a lot. This is important to remember. You don't actually need any expensive kit, or airfare, to experience excitement and wonder. For a toddler walking through a field of long grass, discovering the mini beasts and plants there, is a wild journey in itself: there's a reason, after all, why *We're Going on a Bear Hunt*, about a family walk, is one of the best-loved children's books. And even the moodiest teenager will unfold and relax sitting around a campfire, before sleeping under the stars.

One of the most brilliant adventures I've had with my kids was a late-night walk around a lake near our house. The size of the shimmering white moon, the feeling of the dried grasses moving beneath our palms, and the wild stories we came up with as we walked and talked. It was the simplest journey, but it is a memory that's now an important part of the narrative of our family life.

Even when you do take children far away, on a distant travel adventure, the thing they remember can be surprising. After I took my five-year-old son on a riding safari in South Africa, what excited him most wasn't passing through herds of zebra or giraffe, but the small, strange beauty of the skin of a puff adder he found in the dust near our lodge.

Of course, adventure travel, when it does happen further afield, is a glorious, mind-expanding way to teach children about the world we share, while also encouraging them to walk on that world with greater fortitude, patience, resilience and respect.

Kayaking through rain-drenched fjords in Norway, my then eight-year old and five-year-old learnt that although the natural environment and heavy rain weren't something they could ever control, being prepared and thinking ahead meant the difference

between a wet but exhilarating day and a soggy, cold and uncomfortable one.

Adventure travel also teaches children that with risks come rewards. My six-year-old daughter might have objected to leaving the safe, warm comfort and endless hot chocolate of an Austrian ski lodge at dawn, but once she ventured out into the cold, she was rewarded with a real-life *Frozen* world of virgin snow and glittering blue skies.

If the idea of venturing far afield with children is daunting, remember that it can often be easier than your day-to-day routine at home. Babies, especially new-borns, are surprisingly easy to travel with, since they're so portable, and even toddler tantrums can be easier to manage when not cooped up at home. And the world around – from a bustling market to a busy train ride to a walk through a park – is a natural learning environment, far different to the confines of a classroom.

Sure, there will be arguments and tears along the way and relationships will be tested. But more than anything adventure travel strengthens the family unit. The best week we spent together as a family was cycling down the Danube, almost the length of Austria, from Linz to Vienna, when the children were teens, pre-teens and the youngest was a baby, wobbling along for the journey on the back of a bike. It was a tough, brilliant, exhilarating week, and one which left me with an abiding sense we'd really achieved something as a family.

"You don't actually need any expensive kit, or airfare, to experience excitement and wonder."

Above (left and right): Monkeying around near the Ouzoud Falls, Morocco; spotting elephants in Zimbabwe.
Previous page: Riding camels in the desert

DESERT SAND DUNES
Oman
One of the oldest civilisations on the Arabian Peninsula, Oman is ripe for family adventure: camp in Bedouin tents, take a thrilling desert ride down Jebel Akhdar, go dolphin watching, or explore the dunes at Wahiba Sands and Nizwa, the former capital with a 17th-century fort. Avoid extremely hot July and August; October to April are ideal.
HOW Families World Wide (www.familiesworldwide. co.uk) offers a seven-night trip including one night in a tent.

SAFARI ADVENTURE
Zimbabwe
Introduce your children to safari at Gonarezhou Bush Camp in Zimbabwe, a mobile, tented camp, and at Amalinda Lodge, in the Matobo Hills. Since there are few big cats in this area, children can actually get out of the safari vehicles and explore the area. There's also ancient cave art to discover, and a young explorer programme enabling children to get to grips with safari basics. Best done from September to December.
HOW Mavros Safaris (www. mavrossafaris.com) runs a 10-night trip, with flights to Victoria Falls and within Zimbabwe.

THE ATLAS MOUNTAINS
Morocco
After exploring the maze-like souks of Marrakesh (a great way to introduce children to the magic of Morocco), head into the Atlas Mountains. Take a camel trek up to the Kasbah before travelling through the villages of Timalizen to the land of the Berbers. For thrill-seeking children there's white-water rafting and abseiling. The beachside beauty of Essaouira is a worthwhile stop, especially for the fish markets and watersports. Avoid June to August, the hottest months.
HOW Scott Dunn (www. scottdunn.co.uk) offers seven nights in Morocco.

BE UPLIFTED BY COLOUR

Psychologists have long stated that colour can influence our emotions, and in cultures across the world different hues are used as a way to express. Colour can invigorate and soothe, you just need to know where to look.

HOLI FESTIVAL

India

The most boisterous of Hindu festivals, Holi waves goodbye to winter (usually in March) and welcomes in spring in a rainbow of colours. Hindus celebrate by throwing coloured water and *gulal* (powder) at anyone within range.

HOW The date of Holi changes every year. Mathura in Uttar Pradesh and Mumbai are great places to celebrate.

GRAND PRISMATIC SPRING

USA

Explore the multicoloured mist of this gorgeous pool and its spectacularly vivid rainbow rings of algae in Yellowstone National Park. From above, the spring looks like a giant blue eye weeping exquisite multi-hued tears.

HOW For the most dramatic photos, drive south to Fairy Falls trailhead and walk for 1 mile (1.6km) to the platform.

KEUKENHOF GARDENS

The Netherlands

Botanists, flower-lovers and tourists alike make the pilgrimage each year to one of the world's most famous flower gardens, Keukenhof in Lisse, to ooh and ahh over winding rows of tulips, delicate orchids and snow-like fluttering blossoms.

HOW The park is open for eight weeks in March to May; Lisse is a 40-minute bus journey from Amsterdam.

CHEFCHAOUEN

Morocco

No one really knows why the buildings in this Moroccan mountain town are painted blue. The simple truth is they are very blue, blindingly and beautifully so.

HOW The town is three hours by bus from Tangier, or an hour by grand taxi from Tétouan.

TAKE IT FURTHER

See an awesome
visual spectacle:
Witness a swarm, p112
Tap into your artistic side:
**Unleash your
creativity, p130**

Left: Grand Prismatic Spring, Yellowstone National Park

VOLUNTEER AT HOME

Give some time and maybe some talent to an organisation in your own neighbourhood, and you'll find the cliché about getting out more than you put in rings truer that you can believe.

How can we achieve that elusive state known as 'happiness'? Even the most basic research will reveal that volunteering – helping others – is one sure way to feel good about yourself. And it makes sense in more ways than one. If you're feeling helpless about the problems around you, whether it's homelessness, social isolation or rubbish-strewn waterways, the most satisfying course of action is to take some action. Getting out of your personal bubble can also be a revelation, and can add real balance and perspective to your life. Many local projects cut across social boundaries, and offer helpers profound involvement in their community. Teamwork with other volunteers is a great way to bond and forge friendships, and you can learn new skills or enhance existing ones: there are plenty of community gardens for green-fingered folks; budding bakers can cook for food poverty projects; social butterflies might want to commit to befriending older people; and if your skill is simply that you have time on your hands and can listen, then you can spend a few hours a week at a helpline.

New organisations offering speed volunteering also allow busy people to take part in one-off short projects that fit with their schedule. Devoting even a couple of hours a week to a local project can make a real difference to those around you, and the well-being benefits in all directions may take you by surprise.

LISTENING FOR A HELPLINE
UK

A long-established charity based in the UK and Ireland, Samaritans provides a helpline for people in distress and delivers a crucial service; volunteers are required to be open and non-judgemental.
HOW Volunteers are trained in sympathetic listening, and usually spend three–four hours per week helping; www.samaritans.org.

YOUTH WORK
USA and Australia

Australian organisation AIME (Australian Indigenous Mentoring Experience) aims to support children aged 12–18, building a bridge between school and university for the socially disadvantaged.
HOW University students commit to mentoring young people in group workshop environments; www.aimementoring.com.

Below (left and right):
Volunteers from the Parkholme Supper Club and Growing Communities, London, UK

FEEDING PEOPLE
USA

The USA's largest domestic hunger-relief project relies on volunteer support across a range of projects that connect hungry people with food.
HOW Volunteers can sort and pack produce for food banks, or help out at gardening and educational schemes; www.feedingamerica.org.

EXERCISE AND VOLUNTEERING
UK

A London-based project that certainly has legs, the Good Gym combines fitness with some feel-good community involvement: the volunteers jog in order to help out with local projects.
HOW There are collective opportunities, for example clearing a school garden, or helping older people with domestic tasks; www.goodgym.org.

© HELENA SMITH

How washing up made me happier

"As a travel writer and photographer I've always looked overseas for unusual people and photogenic food, until I realised I could find it all within my own diverse neighbourhood of Hackney in London. While researching a book of recipes sourced from my community, I connected with two homelessness/migrant projects that put feeding people at the heart of their work, and I decided to do a sample stint of volunteering.

I worked one session at North London Action for the Homeless, and I was hooked. I loved the simple shared tasks involved in preparing meals, the rapport with volunteers and service-users, and the feeling each week that I was involved in an endeavour much bigger than myself. Each Monday, I'd turn up in the morning for vegetable prep and table-laying, then I'd welcome service-users as they arrived for lunch. I joined an energetic band of volunteers providing for up to 100 diners, and helped with the epic task of clearing and washing dishes.

Five years on, the cookbook has raised thousands of pounds and I'm still volunteering once a week, plus helping maintain the kitchen garden that provides produce for the meals. I'm fitter (thanks to the gardening), have picked up kitchen skills, enjoy more and deeper friendships, and my attitude to the world around me has altered: I feel more socially aware and thankful for the simple pleasures of home and food."

Helena Smith, travel writer and photographer

TAKE IT FURTHER
Make a difference:
Support a social enterprise, p110
Protect the vulnerable:
Help save an endangered species, p232

HAGGLE

In the sensation-soaked souks and millennia-old markets of the Middle East, Africa, Asia and South America, bartering and the banter that goes with it is an essential experience.

When you come from a culture that perceives pricetags as a fixed part of any purchasing experience, haggling over handicrafts, hats, hammocks and wall hangings in a cacophonous market place can be confronting – but the bartering buzz is very addictive. There's a visceral thrill to be found in the cut-and-thrust of a haggling exchange, probably because it's something humans have done since time immemorial. As an age-old form of human communication, it's an innate skill that awakens in certain environments. A bartering backstory also imbues your travel trinkets with an extra narrative layer, which adds immeasurably to their unique value. Of course, it's just as important to know when not to haggle too hard, and when and where to simply pay the asking price without a fuss. But in most markets, the wily vendors are the experts, and while you might come away with a bargain in your mind, they won't let their wares go for a song.

TAKE IT FURTHER

Be open to romance:
Have a fling, p124
Understand and be understood:
Master a foreign tongue, p260

How to haggle

1. Identify an item you're interested in, and have a good look and feel to assess its quality. Decide whether you really want it (don't waste the vendor's time if you have no intention of buying something), and how much you're willing to pay.

2. Ask the price. Whatever the answer, casually shrug and walk on. Have a good scout around to see how many other vendors are selling a similar product and repeat the step above.

3. Return to your preferred vendor and ask again how much the item is. Suggest you've seen it cheaper elsewhere and make a cheeky offer well below their asking price.

4. Observe the theatrical response, engage your sense of humour and get into some polite haggling. Don't be aggressive or over-demanding. Do be respectful. Keep making counter offers until you arrive at a price close to the one in your head.

5. Be prepared to walk away if you feel something is still overpriced, but be realistic about its value and how much time and labour might have gone into making or procuring it. Remember, though, this process isn't a game for the vendor, it's their livelihood.

6. If you want more than one item (or have a travel buddy who wants something as well), find a vendor selling both and use your combined purchasing power to get a better deal.

MT MERU CURIOS & CRAFTS MARKET

Arusha, Tanzania
Located halfway between Cairo and Cape Town, and situated underneath the incredible cone of Kilimanjaro, Arusha is a vibrant town where you can pick up some superb Maasai materials and handmade crafts at a number of street bazaars – especially the Mt Meru Curios and Crafts Market (aka Maasai Market). Vendors are tourist savvy – bargain hard and halve the asking price.
HOW Find the market on Fire Road. It's open daylight hours, but closes early on Sunday.

KAPALI ÇARŞI

Istanbul, Turkey
Welcoming about 90,000,000 visitors annually, Istanbul's Grand Bazaar is one of the busiest and oldest continuously operating markets in the world, and vendors here will test your haggling skills to the max. The market has been here since the Ottomans conquered Constantinople in the 13th century, and it's a great place to score jewellery, leather goods and carpets.
HOW It's open from 9am to 7pm Monday to Saturday. Arrive by tram, alighting at Beyazıt-Kapalıçarşı.

OTAVALO MARKET

Ecuador
With a dramatic backdrop formed by Andean peaks and Mojanda volcanoes, Otavalo market is a window into pre-Colombian life in the region. Otavaleño stall-holders selling everything from alpaca-wool jumpers and traditional musical instruments to raw food and live animals.
HOW Otavalo is around two hours by bus from Quito. Come on a Saturday, arriving as early as possible.

MARRAKESH

Morocco
Whether you want to bag a carpet, find a fez or take home a tangine, the Red City's ensemble of souks won't disappoint. Fabrics are found in Souk des Teinturiers and Mellah, traditional slippers in Souk Smata, jewellery in Souk des Bijoutiers. For a jumble of treasure and trash, peruse Bab El Khemis flea market. For food, hit Djemaa El Fna.
HOW Souks generally open 9am to 9pm; hours are limited on Friday.

Left: Mounds of spices in the Grand Bazaar, Istanbul

It's good to be bird-brained. Travel in search of rare feathered friends and you're likely to find yourself among the planet's most spectacular, soul-stilling landscapes.

There's a quiet revolution going on among travellers. Where once bungee jumps and diving with sharks enticed globetrotters to wild corners of the world, the travel buzzword du jour is birding. From tree bathing to long-distance hiking, studies prove being out in nature soothes the mind, body and soul. And birding is a meditation for our time: a way to unplug and reconnect with some of the planet's oldest creatures. It's a time to be in quiet solitude yet among the company of like-minded others.

Inside a bird hide, silence settles like a blanket. People move around each other in a careful contactless ritual, to avoid bumping equipment and obscuring lines of sight; there's a hushed effort not to make any sound whatsoever. A hide brings a church-like reverence for place and time, all eyes and ears focused forwards, studying, searching and drinking in the details of nature. And what nature it is.

Birds are the ultimate world travellers, many spending days on the wing, spanning continents in a single journey. They are small but mighty; they sing us to sleep and rouse us awake; they are our planet's ambient soundtrack, its balletic backdrop. From murmurations of starlings over British coastlines to the jewel-like hummingbirds decorating Ecuador's cloud forest, birds are a fine, feathered travel inspiration to us all.

HIDE OUT TO W

ARCHIDONA VALLEY
Málaga Province, Spain
Just half an hour from the Costa del Sol, the Archidona Valley sits right under the avian migratory motorway between Africa and Northern Europe. Here you can spot record numbers of bird species – easily 100 in a day – in a remarkably small area. See griffon vulture, Bonelli's eagle, marsh harrier and, from May to September, hundreds of nesting flamingos.
HOW Birdaytrip (www.birdaytrip.es) runs day trips to Archidona, including a guide.

MANU NATIONAL PARK
Peru
Eco-diverse Manu National Park in the Peruvian cloud forest and Amazon is home to more than 1000 bird species including macaws, parrots, hummingbirds and the distinctive courtship-dancing cock-of-the-rock.
HOW The reserve, located about six hours west of Cusco, has several lodges specialising in wildlife watching and bird safaris, including the Manu Wildlife Center, located right by a macaw clay lick.

TITCHWELL MARSH
Norfolk, UK
The RSPB offers birdwatching lessons for beginners at many of its UK reserves. At Titchwell Marsh in Norfolk, for example, you can see migratory birds as they fly in from the Arctic, bitterns, warblers and black-tailed godwits wading through lagoons, plus, from November to March, thousands of pink-footed geese gathering in their thousands.
HOW The RSPB runs regular birding walks lasting a few hours; see www.rspb.org.uk for the schedule.

ATCH BIRDS

TAKE IT FURTHER
Stand on top
of the world:
Summit a mountain, p152
Be patient, and rewarded
by a rare animal encounter:
Watch and wait, p182

© DAVID DORAN

LEAVE YOUR SMARTPHONE AT HOME

Many of us will readily admit to spending too much time glued to our smartphones, but can they really detract from a sense of place? And if so, how can we better engage our senses, tune into nature and connect with others to become more mindful, conscientious travellers?

Sure, we know it's hard. In a switched-on world, being constantly connected via your smartphone and social media has become the norm. We can instantly ping travel photos over Whatsapp, get multiple 'likes' for envy-inducing snaps of gorgeous beaches on Instagram, impress followers with selfies of us diving in tropical reefs or dangling off cliff edges. But aren't we missing a trick? Unplugging can be as easy as taking a paper map out when hiking or ditching Google recommendations in favour of asking a local.

When we obsess over recording every second of our travels with our smartphones, we are in danger of letting the true spirit of a place pass us by. And surely finding that spirit is why we travel in the first place. So it's important to know when to leave the smartphone behind. Trust us, it can be liberating.

The effects of smartphone overuse are well documented, with studies showing higher levels of loneliness, depression and anxiety. When you travel, this might translate as interacting online rather than face-to-face, thereby missing out on cultural insights, new friendships and one-of-a-kind wildlife encounters. And there are more creative ways to remember your travels, should you so wish: from sketching to letter writing or keeping a travel diary – telling rather than showing.

© MEKOSHA

BEGINNER: PUT YOUR PHONE AWAY

If you walked into a hostel a decade ago, you'd have found a sociable group of travellers trading tales and tips. You went to cafes to strike up conversations, markets to haggle and try unfamiliar foods, and stopped people in the street to ask for directions. Then came the smartphone. Switching it off now can rewind time and help you connect with others.
HOW Though a few cafes out there are now laptop-free zones – Dough Lover (www.doughlover.com) in Brighton, England, to name just one – hotels have been slower to embrace the trend. But you can exert control by allotting yourself a specific time to check your phone each day, and otherwise disconnecting.

ADVANCED: DISCONNECT COMPLETELY

Taking a break from technology can have a calming effect and help us reconnect with our physical and spiritual selves. Ditch the phone and book into a tech-free retreat for a day, week or month – be it a yoga and meditation retreat in Thailand, an Ayurveda escape in India, or a stress-busting spa by the sea. The longer the stay, the more powerful the effect.
HOW Kerala's Ayurveda retreats set a glowing example when it comes to non-tech environments. Boutique-style, riverside Mekosha (www.mekosha. com) near Thiruvannthapuram (Trivandrum) has banned wi-fi in the rooms and dining areas, and phones are only permitted in a designated area.

ADVANCED: GO REMOTE

One way to kick-start a digital detox is to remove yourself completely from the source by travelling to a remote place with no signal or wi-fi, be it an island in Indonesia, a national park in the Australian outback, or a multiday hike in the Alps. In wild places, it's easier to engage the senses, whether wildlife spotting, sleeping under the stars or foraging – anything, essentially, that deepens your connection with nature.
HOW Embark on a multiday hike where the signal is sketchy: starting and ending in Chamonix, the 106 mile (170km), 11-day Tour du Mont Blanc is an epic choice in the Alps, dipping into France, Switzerland and Italy, with hut accommodation offering shared meals. Book ahead in summer.

A screen love affair

Most of us spend more time on our phones than we realise (or care to admit), with the average user checking updates every 15 minutes, but when do we cross the fine line to addiction? If recent neuroscience studies are to be believed, we're not so much addicted to smartphones as we are to the human impulse to interact with others, making us 'hypersocial' rather than 'antisocial'. To not be online leaves us with raging FOMO (fear of missing out).

But the dangers of overuse are becoming apparent, with everything from low mood and anxiety (driven by the reward system) to decreased focus and physiological symptoms like increased heart rate and blood pressure cited. The solution, it seems, is restraint, with neuroscientists suggesting ways to curb phone addiction and reinforce better digital habits, from turning off push notifications and allotting time slots for checking social media to banning phones from the bedroom – all simple means of regaining control.

TAKE IT FURTHER
Be present and inspired:
Engage all your senses, p188
Practise with the best :
Meditate with masters p216

Left: Room at Mekosha, Kerala, India

LIVE WITH LOCALS

Bedding down with a family offers a revealing insight into local life and culture and puts your tourism dollars straight into the hands of your hosts.

Opting for a homestay is one of the most authentic ways to experience local culture and hospitality in any destination. It allows you to actively participate in the local way of life and maybe even learn a new skill, such as how to prepare a traditional meal.

Homestays aren't designed to be like hotel stays. In fact, staying with a local family can be challenging, particularly when there's a language barrier or when local customs (not to mention comfort levels) differ greatly from those of your own country. But the rewards are rich. Not only do responsibly run homestay experiences ensure your tourist dollars directly benefit your hosts, the experience offers a window into local culture that no other accommodation can offer. By embracing the experience, you're bound to depart with a deeper knowledge of local life in your destination, and perhaps even a friend or two for life.

Many socially responsible small group tour operators work with local NGOs – or directly with communities – to develop meaningful homestay experiences. Homestay booking websites are becoming increasingly common, but the experiences can be variable.

KATHMANDU VALLEY
Nepal
Whip up traditional Nepalese dumplings and help with the rice harvest during a homestay experience arranged by Rickshaw Travel. It works with a community project in Panauti, in the Kathmandu Valley, designed to empower local women.
HOW Rickshaw Travel (www.rickshaw travel.co.uk) offers a one-night homestay programme including meals.

PHANG NGA BAY
Thailand
On the island of Ko Yao Noi in Phang Nga Bay, the community run a homestay project where visitors live (and eat!) like a local for a night or two. Part of the fee goes to finance environmental and educational projects.
HOW One-night homestays, including transfers from Phuket, meals and activities, can be arranged via www.facebook.com/kohyaoonoihomestay.

Below (left and right): Meas family homestay in Cambodia

IN AND AROUND CUSCO
Peru
Intrepid Travel run homestays in a farming community north of Cusco. Freshly ground coffee, home-grown cacao and hearty meals are on the menu.
HOW Try the 10-day Peru Real Food Adventure by Intrepid Travel (www.intrepidtravel.com).

MEAS FAMILY HOMESTAY
Cambodia
This popular and friendly family homestay has 12 rooms in a spacious compound located between Angk Tasaom and Takeo. It's a gorgeous place to ride a bike around the rice paddies, and the homestay provides complimentary wheels.
HOW Book your place at ww.cambodianhomestay.com.

Cooking up a feast in Jordan

"As I watched the whitewashed high-rises of Amman dissolve into the undulating olive groves of rural Jordan out of the taxi window, I started to feel a bit nervous. Was signing up for a homestay experience in a country I'd just arrived in – knowing a mere few words of Arabic – a little ambitious? But my anxiety dissipated when my driver pulled up in front of a turquoise gate in the verdant highlands of north Jordan, where my host Mahmoud was waiting for me with a wide, disarming smile. Feeling instantly at ease, I settled in for an afternoon of gossiping with his wife Maysoon, about everything from politics to marriage to what it's like to be a woman in contemporary Jordan, while we sipped cardamom-infused coffee and stuffed vine leaves for the evening meal. After helping the couple's children set the 'table' (a tablecloth spread on the driveway, as it was too hot to dine inside), we sat down to what would be one of the most delicious meals I'd eat all year.

Experiencing this warm Jordanian hospitality was one of the highlights of my visit to the country, and every time I smell spiced coffee, I smile at the memory."

Middle Eastern operator Engaging Cultures (www.engagingcultures.com) offers a homestay experience with a local family in Orjan, a Jordanian village an hour's drive from Amman.

Sarah Reid, travel writer

TAKE IT FURTHER
Get to know a minority community:
Live on the fringe, p190
Commit long-term:
Give a year of your life for others, p280

FIND PEACE AMONG THE TREES

There's no solace greater than that which the forest offers. Whether walking alone in leaf-strewn autumn woodlands, climbing high into the branches of an ancient oak, or watching light rake through freshly formed buds on a spring morning, trees calm us down, make us mindful and keep us grounded.

'And into the forest I go, to lose my mind and find my soul,' said Scottish-American naturalist John Muir, neatly summing up the effect trees can have on the psyche. The forest, never entirely silent yet always peaceful, is more protective than the open sea with its canopy and more approachable than the high mountains, despite often being portrayed in a sinister light in folk tales, with its sly foxes, big-bad wolves, thieving bandits and wicked witches. The forest is a dark place. Or is it?

In the 1980s, the Japanese cottoned on to *shinrin-yoku*, now better known as 'forest bathing'. Get it right and this health-giving practice can have powerful benefits: calming the nervous system, boosting the immune system, slowing the heart rate and relieving all manner of stress. In the Western world, a similar but little-used term is nemophilist: one who is fond of forests, a haunter of the woods.

Step-by-step: Forest bathing

1. Pick your location. Make sure it's quiet, and as far removed from distractions and traffic as possible. Turn your phone off and preferably go alone.

2. Start small. Select an area you can become intimately acquainted with – perhaps just one tree. Be still and spend at least 20 minutes in one place.

3. Pay attention to the details – the patterns on a leaf, the rough textures of bark, the smell of pinesap. Be fully present and notice things on a minute scale.

4. Engage all your senses: look out for wildlife (from fleeting deer to tiny insects), feel the forest floor underfoot (remove your shoes if you so wish).

5. Pay close attention to your breathing and imagine the root system of the forest extending through your feet, pulling them into the earth, for a grounding effect.

TAKE IT FURTHER

Connect with nature:
Spend a night in the jungle, p128
Feel calm on another level:
Meditate with masters, p216

THE BLACK FOREST
Germany

One of Europe's loveliest forests spreads its thick coniferous blanket over the valleys of southwest Germany. It's special year-round, but particularly on golden summer evenings (June to August), crisp autumn days (especially September), and when frosted with snow.
HOW The closest airport, and gateway, is in Baden-Baden; you'll need a car to explore properly.

ARASHIYAMA BAMBOO GROVE
Japan

Japan is the birthplace of 'forest bathing' as a concept. While the options for *shinrin-yoku* are boundless, one fine spot close to Kyoto is Arashiyama Bamboo Grove, where you can walk in silent wonder among towering bamboo stems.
HOW The grove is located 6 miles (10km) west of the city centre. Take the commuter train from Kyoto Station to Saga-Arashiyama Station.

GREEN MOUNTAIN NATIONAL FOREST
USA

Time your visit to this vast forested, mountainous swath of Vermont for autumn (mid-September to October) when the maple, poplar and birch kindle into a riot of scarlets, rust reds, golds and ambers. You'll also find looking-glass lakes and moose, black bears and white-tailed deer.
HOW The airport in Albany is a two-hour drive from the Green Mountain National Forest; www.fs.usda.gov/gmfl.

GOBLIN FOREST
New Zealand

The volcano of Mt Taranaki watches over eerie Goblin Forest in East Egmont and walking here is a real glimpse of Middle Earth. Mist often hangs between the gnarly, lichen-swathed kamahi trees, thick mosses and ferns, while natural plunge pools invite contemplative moments and cooling swims.
HOW It's roughly a five-hour drive south of Auckland. Good bases include Konini Lodge and The Camphouse.

LEARN ABOUT THE DAF

A rarely considered but critical aspect of travelling is bearing witness: seeing the sights and hearing the stories of the tragedies and the atrocities that have occurred in a place, in order that they might never be forgotten.

*T*ravelling is supposed to be fun, right? Well, sure, most of our trips revolve around personal enjoyment – dinners, beaches, photo-worthy landscapes. But consider dedicating at least a portion of your sightseeing to something a bit less comfortable: visiting sites of great historical sadness. There are two important reasons why you should. First, there's the aspect of bearing witness. As the ever-repeated quote by Spanish philosopher George Santayana goes, 'those who cannot remember the past are condemned to repeat it'. And there are a great many things we don't want to repeat. The Holocaust, the Cultural Revolution, the violence and segregation of the Jim Crow South, the Rwandan genocide. Evidence suggests that we do, as a society, forget even the most barbaric events over time. A 2018 survey in the USA showed millennials have a dramatically worse understanding of the Holocaust

than their elders – only 66% knew what Auschwitz was. A remedy? To see the places where these events took place. No one who visits a concentration camp or a killing field will ever forget or doubt the terrible reality of what happened there.

The second reason is more personal: knowing about great evil can spur us to do good. We may know, intellectually, that landmines are a terrible thing, but visiting a landmine museum and learning stories of their victims moves us in a different way. Maybe then we decide to donate money to anti-landmine causes or to volunteer with injured children. Or maybe we just become a bit kinder.

That said, try to avoid sites that treat atrocities as displays for voyeurism rather than opportunities for education, and ask whether the victims of the incidents in question are benefiting from your visit.

KER SIDE OF HISTORY

CHOEUNG EK KILLING FIELD
Cambodia

This peaceful longan orchard turned unspeakably bloody in the late 1970s, when 17,000 citizens were transported here for execution. Their crimes? Anything from being well-educated to being part of an ethnic minority to practising Buddhism. Some of the mass graves have been exhumed; there are 8000 human skulls on display in the Memorial Stupa. But many of the bodies still lie half-buried. Seeing a femur, for instance, protruding from the ground is an indelibly – and intentionally – disturbing experience.
HOW Visit by tuk-tuk from Phnom Penh. Admission includes an excellent audio tour.

THE NINTH FORT
Lithuania

Approximately 10,000 Jewish men, women and children were slaughtered by Nazis in a single day in this 19th-century fort in the Lithuanian town of Kaunas. The fort is now a museum and monument to the Holocaust and to the victims of the Kaunas Massacre, an event with far less name recognition than other WWII atrocities.
HOW The museum (www.9fortomuziejus.lt) can be reached by public transport from Kaunas. Guided tours cover different aspects of the site.

NATIONAL CIVIL RIGHTS MUSEUM
USA

On 4 April 1968, Martin Luther King Jr was shot and killed while standing on the balcony of the Lorraine Motel in Memphis. The motel is now part of a museum dedicated to understanding the history of the Civil Rights movement and the ongoing equality struggles of African-Americans. Interactive exhibits are shattering: sit next to Rosa Parks on a bus or listen to racist jeers as you sit at an all-white lunch counter.
HOW Visit the museum at 450 Mulberry St, Memphis. www.civilrightsmuseum.org.

Below: Harrowing display of human skulls in the Memorial Stupa, Choeung Ek

TAKE IT FURTHER

Explore your own personal history:
Retrace your roots, p154
Recalibrate your viewpoint:
Challenge your perceptions, p210

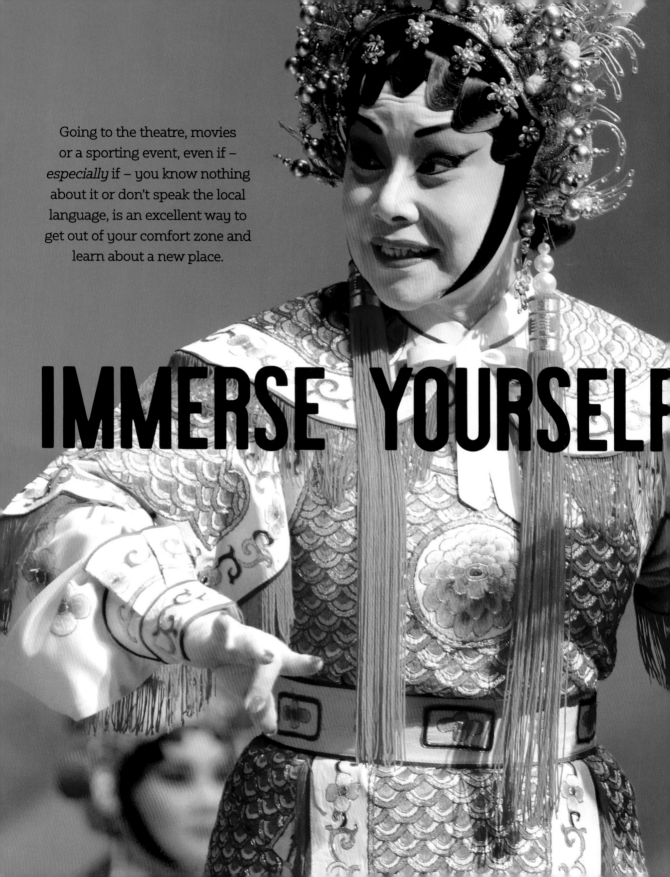

Going to the theatre, movies or a sporting event, even if – *especially* if – you know nothing about it or don't speak the local language, is an excellent way to get out of your comfort zone and learn about a new place.

IMMERSE YOURSELF

Why on earth would you attend a play in a place where you don't speak the language? What could motivate you to attend a sporting event if you would never consider doing so at home?

Well, think about this: performances and events – sports, theatre, dance, music, movies – are national pastimes, and as such offer a unique window into traditions and values of our destinations. You can learn a lot from watching, even if you don't understand a fraction of what's going on. You'll see how locals behave – are they quiet or loud? Do they applaud or snap? Eat snacks and chatter, or watch with reverence? Paint their faces and wear crazy wigs, or come formally dressed? Watch how people interact with each other and with the performers to get a sense for social rules and communication styles. Performances themselves can speak across language barriers as well. You don't need to understand Hindi to appreciate the passionate love songs and musical numbers of a Bollywood film. You don't need to know the rules of football (soccer) to get a rush out of cheering along at a game in São Paulo.

Events are often good ways to meet locals as well. People love nothing more than to talk about their passions, so sitting next to a cricket fanatic at the Melbourne Cricket Ground and asking for a quick explanation of bowling procedures is surefire way to spark a conversation.

N LOCAL CULTURE

TAKE IT FURTHER

Heed expert advice:
Learn from indigenous cultures, p120

Push yourself:
Get out of your comfort zone, p180

Left: Cantonese Opera, Hong Kong

LUCHA LIBRE
Guadalajara, Mexico
The Tuesday night fights at the iconic, slightly decrepit Arena Coliseo in central Guadalajara is a window into a certain corner of the Mexican soul. Masked *luchadores*, divided into *técnicos* (good guys) and *rudos* (bad guys), fly at each other in a series of highly choreographed moves, as crowds roar, jeer and chat noisily, all while snacking on doughnuts and downing *micheladas*. It's rowdy, no-holds-barred fun.
HOW Fights at Arena Coliseo de Guadalajara take place at 8.30pm Tuesday and 6pm Sunday; tickets can be bought at the door. www.cmll.com. Medrano 67, Guadalajara, Jalisco 44450.

CANTONESE OPERA
Hong Kong, China
With roots stretching back nearly 1000 years, Cantonese opera is one of the world's oldest art forms. You can find it all over Southern China, though Hong Kong has some especially fun historic theatre venues. Performances combine singing with martial arts and acrobatics, and elaborate costumes have complex symbolism; read up beforehand to better understand.
HOW Catch a show at the vintage Sunbeam Theatre in Hong Kong's North Point; be aware that performances can easily last three hours. See http://sunbeamtheatre. com/hk for performance schedules.423 King's Rd, Kiu Fai Mansion, North Point.

ANCIENT GREEK THEATRE
Athens, Greece
Watch a play almost as old as civilisation itself in Greece, the birthplace of Western theatre. Athens has several ancient venues worth a trip in themselves, including the Odeon of Herodes Atticus, dating from AD 161, on the slope of the Acropolis. Tragedies such as Euripides' *Electra* and comedies like Aristophanes' *Plutus* shed light on problems that we still tangle with millennia later – the randomness of chance, the unequal distribution of wealth, the vagaries of love.
HOW Catch a summer show at the Athens & Epidaurus Festival (www. greekfestival.gr).

FIND YOUR TRIBE IN A FOREIGN CITY

Whether you're a fan of jazz jams, the tufted titmouse or triple IPAs, your people are out there. Joining them can make a foreign land feel much more familiar.

*A*rriving in a new city can be unnerving – the language sounds odd, the traffic is crazy, it smells weird – and it's easy to be deterred from going out and immersing in the scene.

This is where your tribe comes in. Finding others who share your interests can be a great motivator to do something you might not otherwise do. Let's say you're in Dakar. It's chaotic, overwhelming and intimidating to explore. Then again, you love running. You search online, find a public running club and, voilà, next afternoon you're out with a group of locals hurtling through streets you'd never experience otherwise. As a bonus, the runners go for a drink afterwards, so you hang out in a cool neighbourhood bar with new acquaintances and get tips galore on city sights.

Technology makes it easy to locate your group. Fellow knitters in Buenos Aires? Bluegrass music fans in Hyderabad? Arsenal supporters in Chicago? They await your presence. Check Meetup (www.meetup.com), a tool that event organisers use, or search for businesses that are well known in their field (beer enthusiasts can seek out esteemed beer bars, jazz fans popular clubs).

Escapades with your tribe often end up being a trip highlight, and it's not just because you're gathering with like-minded types. The activity itself plays a role. Research shows that familiar things make us feel comfortable. If ukulele strumming is your jam, then ukulele strumming in Abu Dhabi makes the city that much more agreeable.

NEW YORK CITY

1

Times Square •   • Grand Central

Penn Station ⓡ

MANHATTAN

Hudson River

East River

QUEENS

4

2

• World Trade Center

• Brooklyn Bridge

3

BROOKLYN

Left: Join fellow twitchers in Central Park, NYC

❶ IF YOU LIKE... BIRDWATCHING

Go to... Central Park
Join the folks looking up trees and blurting out names such as, 'Palm warbler! White-breasted nuthatch! Red-tailed hawk'. Guided walks take place in the mornings at different locations around the park. **See www.birdingbob. com/birdwalks.**

❷ IF YOU LIKE...BEER

Go to... Proletariat
The cognoscenti of NYC's beer world pack this tiny bar that pours a changing line-up of uncommon brews. Beer buffs will feel right at home discussing everything from hop varieties to triple IPAs with the knowledgable bartenders. **Check out www.proletariat-ny.com.**

❸ IF YOU LIKE... STREET ART

Go to... Bushwick
In Brooklyn's gritty Bushwick district, local businesses offer up their walls and then a who's who of artists paint them. The murals change regularly and are mainly along Jefferson St and Troutman St between Cypress and Knickerbocker Aves. **See www.instagram. com/thebushwickcollective.**

❹ IF YOU LIKE...CHESS

Go to... Washington Square Park
Test your wits with one of the speed chess players who sit at the tables dotting the park's southwestern corner. The skill level is usually high. It'll cost you a handful of dollars per game to go up against one of the trash-talking hustlers. **See www.chessnyc.com.**

HONG KONG

KOWLOON

Hong Kong West Kowloon Ⓜ

Star Ferry Pier

Victoria Harbour

Hong Kong–Macau Ferry

❹

❶

Ⓜ Hong Kong

❸

Ⓜ Central

Peak Tram

• Victoria Peak

❷

HONG KONG ISLAND

❶ IF YOU LIKE... TO MEDITATE

Go to... Samadhi Training Centre for the Soul
At ground level there's a meditation space that welcomes drop-in peace-seekers. Stoke your consciousness with like-minded types in the soothing, space. www.facebook.com/samadhicentre.

❷ IF YOU LIKE... QUIDDITCH

Go to... various parks
Calling all Harry Potter fans. Grab your broomstick and brush up on the lingo so you know your quaffle from your snitch and bludger. Then get ready to do battle in this rugby-meets-dodgeball sport. Teams face off every Saturday. www.facebook.com/hkquidditch.

❸ IF YOU LIKE... JAZZ

Go to... Peel Fresco
The charming club has live jazz six nights a week, with local and overseas acts on an intimate stage close enough for listeners to chink glasses with musicians. Tuesday is open-mic night, Sunday a jazz jam of locals. www.peel-fresco.com.

❹ IF YOU LIKE... GIN

Go to... Ping Pong Gintoneria
Artsy cocktail aficionados descend into the cavernous former ping-pong hall to swig gin. The neon-illuminated bar stocks more than 50 types from across the globe, served in a variety of tipples both classic and creative. www.pingpong129.com.

LONDON

❶ IF YOU LIKE... FASHION AND STREET FOOD

Go to... Broadway Market
Artisanal nibbles, arty knick-knacks, books and vintage clothing fill the stalls at the city's hippest market in Hackney, Saturdays 9am to 5pm. www.broadwaymarket.co.uk.

❷ IF YOU LIKE... SWING DANCING

Go to... Sway Bar
Bust out your best Lindy hop and jitterbug moves with the London Swing Dance Society on Monday nights at Sway Bar in Covent Garden, plus the odd foray to other venues. www.swingdanceuk.com.

❸ IF YOU LIKE... CRAFTING

Go to... London Craft Club
If your hands itch to bead jewellery or sew a purse, the Craft Club will sort you out. It holds free craft meet-ups at its studio, as well as classes such as weaving and embroidery. www.londoncraftclub.co.uk.

❹ IF YOU LIKE... BAKING

Go to... E5 Bakehouse
This organic bakery has a reputation for its sourdough bread made with locally sourced ingredients, and DIY enthusiasts laud its weekly classes on bread, jam and pickle making. e5bakehouse.com.

MELBOURNE

❶ IF YOU LIKE... MOVIES

Go to... Sun Theatre
Boutique art-deco glory is on offer at this golden-age cinema that screens both art-house and popular films. A movie club meets regularly to watch and dissect films here. www.meetup.com/Sun-Theatre-Movie-Fans.

❷ IF YOU LIKE... RUNNING

Go to... Albert Park
Every Saturday at 8am, more than 100 runners gather for a free 5km. Everyone signs up in advance to get a barcode, which volunteers scan and use to provide timed results. www.parkrun.com.au/albert-melbourne.

❸ IF YOU LIKE... SLOW FOOD AND ART

Go to... Abbotsford Convent
This collection of ecclesiastical architecture is now home to a thriving arts community of galleries, studios and eateries, with a Slow Food Market every fourth Saturday. www.abbotsfordconvent.com.au.

❹ IF YOU LIKE... DRAWING COMICS

Go to... Squishface Studio
The studio's cartoonists host a drawing night the first Wednesday of the month. A love of comics and willingness to put pen to paper are all that's needed to attend. www.squishfacestudio.com.

The benefits of making things with your hands are well established. Studies show it can reduce depression and anxiety, decrease stress, soothe pain and foster creativity. Unfortunately for many of us, our hands spend most of the day typing away at keyboards rather than producing tangible physical work. Your next trip could be the opportunity to change that. Dedicate some time, whether a morning or a month, to learning a local craft. This could mean taking a short, traveller-oriented workshop offered by a town's tourism department. Or it could mean going deeper, signing up for a longer-term course mostly attended by locals, perhaps even taught in a foreign language.

The rewards for you go well beyond stress-reduction. You can make friends – there's nothing like spending an afternoon together sanding a boat hull or glazing pottery to turn strangers into confidantes. And crafts are often a window into a place's history and values. *Ikebana*, the Japanese art of flower arranging, speaks to the Shinto belief about the divinity of all things, so reverently is each single blossom treated. Italian pasta making shows the deep respect the culture has for tradition and food. Taking a leathercraft class in Morocco will teach you the history of the country's storied leather-tanning industry. So roll up your sleeves and plunge in.

LEARN A CRAFT

Why buy a souvenir when you can make one? Taking a class from an artisan teaches you new skills, lets you work with your hands, and gives you the opportunity to interact with the local community on an intimate level.

BOAT BUILDING
Maine, USA
Just visiting the WoodenBoat School in tiny Brooklin, Maine, where it's all pine forest and crashing grey sea, is worth the trip. Commit to a week to learn the basics of constructing a canoe or skiff from scratch. Learn an exotic new language – plank-on-frame, plywood-epoxy – while sanding and sawing in the workshops.
HOW The on-site accommodation and dining hall at the two-week fundamentals course promotes serious bonding; www.thewoodenboat school.com.

BOOKBINDING
Worcestershire, UK
Have you always loved the smell of a used bookstore? Found that ebooks lack soul? Learn to rescue written treasures with a bookbinding course at Green's Books, a book and paper conservation studio in Worcestershire, UK. The owner has worked at such venerable institutions as Oxford's Bodleian and the British Library. He'll guide you through lost arts like springback binding and leather paring.
HOW Courses last from a day to a week; see www.greensbooks.co.uk.

BATIK DYEING
Bali, Indonesia
Nobody leaves Bali without buying some kind of batik – a textile made with a traditional Indonesian method of wax-resist dyeing. But instead of buying a sarong or tablecloth, learn to make your own, with a workshop at Nyoman Warta Batik. Nyoman is a well-known batik artist based in the upland town of Ubud. Three-hour classes in his laid-back studio move from sketching to wax application to dyeing, all while learning about the history of the art.
HOW For information on the class schedule see nyomanwarta.com.

TAKE IT FURTHER

Exercise your
artistic temperament:
Unleash your creativity, p130
Express yourself
in words:
Write a travel blog, p158

TAKE A PLUNGE IN OPEN WATER

Wild, wet and potentially treacherous, ocean
swimming had never appealed to **Adam Skolnick**,
until a sprained back lured him back to the water.
Beyond the waves, he discovered the cold and
curative powers of the big blue.

*O*n the 30-minute drive up the coast to the beach, I was nervous as hell. I'd tried this once before, swimming well beyond the waves, about a quarter mile offshore, and it had not gone well. It had been an afternoon in early May, and the Pacific Ocean was a hypothermic 56°F (13°C). I wore no wetsuit. My goggles leaked. Within minutes, I panicked and swam desperately for shore.

Two years later, I hadn't forgotten that first encounter so much as suppressed it. I'd sprained my back severely a few months prior and could no longer run to stay fit. A massage therapist suggested I get into the pool, and from the minute I hit the water I knew I'd found something. Without the burden of gravity, my body relaxed. I could flow and build strength at the same time. Then I saw a loose Band-Aid float beneath me and knew it was time to test myself in the Pacific yet again.

As far as aerobic sports go, swimming offers the most cardiovascular benefits with the least abuse on your joints, but when you slip the safe confines of contained water you open yourself to additional mental, emotional and physical benefits. Especially if the water you find is a little bit cold. Swimming in cold water increases your metabolism and blood flow, and with that increased circulation comes a reduction in inflammation. It spurs healing. Then there's the mental challenge of it. Yes, it can be scary to swim out to deep water, but the confidence and self-reliance that comes with that cannot be overstated. You will learn how to be calm in the chaos and figure out when to go hard and when to surrender. In other words, it's great training for becoming a sane adult. Plus, at its best, open-water swimming is an absolute joy.

It was a beautiful summer day in August when I arrived for my second attempt. The sea was calm and about 10 degrees warmer. This time I had brand-new goggles and zero hesitation or terror. From the moment I ducked under the waves and into the clear blue I was hooked. I saw more than 50 rays swimming along the ocean floor that day. It helped that visibility was terrific, but I knew that wouldn't always be the case. Still, I kept coming, two, three, then four times a week. Some days the water was stormy and dark. Others it was calm and clear.

At first, I was happy to tag along and learn from more experienced friends. Now, some six years later, I routinely free-dive beyond 50ft (15m), duck under overhangs to spy lobster, happily swirl alongside friendly sea lions. I've had full-grown grey whales swim directly under me. But it's not the external thrills I think about when I invite friends, old and new, to join us at the reef. It's the soul satisfaction that seeps into my system from the ocean itself.

I remember one day early on, while we were still on the beach suiting up, we saw a 70-year-old woman rise up out of the shallows, her smile ear to ear.

'What did you see out there?' my friend John asked.

'Nothing. The visibility was awful,' she said, smiling.

'How was the temp?'

'Oh, it was freezing. It must have been an upwelling overnight.'

'Really?' I said with a hint of dread. 'But how was it?'

'Oh,' she said, looking back towards the water, shaking out her hair then turning back, 'it was magnificent!'

I knew exactly what she meant.

> "When you slip the safe confines of contained water you open yourself to additional mental, emotional and physical benefits."

LA JOLLA COVE
San Diego, USA
At San Diego's northern end is a sheltered bay that lures open-water swimmers every day. The cove and its kelp beds are home to harmless leopard sharks and sea lions. All you need is a pair of goggles, a swimsuit and a mask, and you're good to go. Head here August to October for the best temperatures and visibility.
HOW Fly into San Diego and drive north on Interstate 5, exit La Jolla Pkwy.

Below: Californian sea lion off Baja California.
Previous page: Swimming over a coral reef

GARDNER BAY
Galápagos Islands, Ecuador
No trip to the Galápagos Islands – an evolutionary marvel, home to endemic species you can see nowhere else – is complete without a snorkel in this cool, crystalline bay where you'll spy sea lions, green turtles and marine iguanas.
HOW Nature Galapagos (naturegalapagos.com) offers day trips. Flights to Isla San Cristóbal, where Nature Galapagos is based, depart Quito daily.

NINGALOO REEF
Australia
This beautiful white-sand beach in Western Australia is like Burning Man for charismatic sea creatures – everybody is welcome. Whale sharks gather from mid-March to mid-September; humpback whales come from June to October. Turtles, dolphins and dugongs hang-out year-round.
HOW Ningaloo Whalesharks (www.ningaloowhalesharks. com) offers whale shark and humpback whale swimming tours, as well as dive trips around the reef. Fly here from Perth then drive 72 miles (116km) to the reef.

TAKE IT FURTHER
Get a head for heights:
Do look down, p134
Feel as free as a bird:
Fly, p256

SEE THE MAGNITUDE OF THE EARTH'S POWER

Seek those hotspots where the earth's energy gathers, rumbles and explodes with beautiful ferocity. Where it's not the mere ooze of glowing lava nor the dynamic blast of a geyser that dazzles, it's the evolutionary context that dwarfs our silly human frailties and shocks us into silent awe.

THE BIG ISLAND
Hawaii, USA
The Big Island made headlines in 2018 when lava burst through asphalt highways. Peep into the crater at Hawaii Volcanoes National Park then watch lava flow on a night cruise.
HOW Book a cruise at Lava Ocean Tours (www.seelava.com).

YELLOWSTONE NATIONAL PARK
USA
This is a geothermal showground with more than 10,000 geysers and hot springs. Old Faithful, which has erupted at least every couple of hours since 2000, draws the biggest crowds. Skip the summer rush and go from October to March.
HOW Fly into Jackson, Wyoming, a 2.5-hour drive from the park gates.

JIGOKUDANI MONKEY PARK
Japan
In the snowy mountains around Nagano, lava courses through the earth, heating up groundwater, which bursts to the surface in a series of *onsen* (hot springs). At Jigokudani Monkey Park, snow monkeys take advantage of this, descending from the mountains to warm up in their very own *onsen*. Visit when there's snow on the mountain (December to February).
HOW Take the Hokuriku Shinkansen from Tokyo. See http://en.jigokudani-yaenkoen.co.jp.

HUANGLONG NATIONAL PARK
China
Huanglong National Park in Central China has intact forests, snow-capped mountains, waterfalls and colourful lakes that are the result of rich calcium deposits swirling in water. Geothermal hot springs burst through the earth's crust throughout the park.
HOW Fly into Jiuzhai Huanglong Airport, 27 miles (43km) from the park.

TAKE IT FURTHER
Go where the wild things are:
Meet the planet's giants, p136
Seek the rare and the special:
Witness a miracle of nature, p282

© JUSTIN REZNICK | GETTY IMAGES

Left: Lava flowing into the sea in Hawaii

FEEL
THE RUSH

Stand with your toes or skis on the precipice of some steep drop and the chemical process will have already begun. A slow drip of the hormone adrenaline will begin coursing through your veins. Your pulse will quicken, sweat will trickle.

The blood vessels in your extremities will contract, sending oxygen-rich blood towards your core, flooding your organs and dilating your air passages to collect even more nourishing O_2. This is the beginning of the fight-or-flight response, and adrenaline is its active ingredient. It provides the turboboost to run like hell or fight like an MMA warrior, if or when you find yourself cornered or threatened. It's biology at work, your survival instinct dialled way up high. And it's autonomic, meaning it's not something you can control. But that doesn't mean you can't ride it.

Adventure sports were built on adrenaline. Downhill skiing, snowboarding, surfing, sky diving and rock climbing are fun on their own, but when the stakes are high and the danger slightly more immediate, the rush of it all delivers a natural high almost impossible to match. Add a shot of pristine nature and an excellent photo op and you've just won the day.

DIVING AMONG THE SARDINE RUN
South Africa

Every year from May to July, vast, swirling shoals of sardines rise up from the depths and migrate to southern South Africa and travel from the temperate waters northeast to the subtropical seas off the country's Wild Coast. As they move, they attract a feeding frenzy of stampeding common dolphins and diving sea birds, as well as sharks and the occasional whale. Imagine an episode of *Blue Planet* – now insert yourself with a scuba rig.

HOW South African dive outfitter Blue Wilderness (bluewilderness.co.za) offers week-long dive trips out of East London, South Africa.

SKIING AND SNOWBOARDING
Whistler Blackcomb, Canada

Whistler Blackcomb, a beloved ski resort just northeast of Vancouver, offers thousands of feet of downhill runs and a sick terrain park. Skiers and snowboarders of all levels shred each winter, while the downhill mountain biking is special in the summer, as is the nearby white-water rafting. Whistler is a family-sized, year-round adrenaline rush.

HOW For those who want a bit more bump and have the skills for it, try Highest Level Terrain Park at Whistler Blackcomb (www.whistlerblackcomb.com). If you want to slowly make your way into more challenging terrain, Big Easy Terrain Garden offers smaller rails, rollers and mini-hits.

CANYONING
Queenstown, New Zealand

Don a wetsuit and helmet, and hike through rushing creeks, scramble up boulders, jump off cliffs into pools and swim a little, too. Queenstown is one of the best places in the world to go canyoning (aka gorge walking), plus it's a great ski town with other summer thrills like heli-hiking and white-water rafting.

HOW Canyoning Queenstown (www.canyoning.co.nz) has been leading guided canyoning adventures since 1998; it offers half-day and full-day trips to Queenstown and Routeburn Canyons.

Left: Skiing at Whistler Blackcomb, Canada

A very brief history of adventure sports

Vikings invented skis as an act of survival, to hunt reindeer 5000 years ago. The evolution of the wooden ski and the switch to aluminium in the 1920s sent alpine racers zooming downslope. For ancient Hawaiians, surfing was a creative and spiritual act. It wasn't commoditised until Waikiki became a tourist destination where Jack London surfed. He wrote about it in 1907. By then, two Hawaiian princes had brought boards to California, where skateboards were invented in the 1940s because surfers needed something to do when there were no waves. Snowboards first appeared about 20 years later and were embraced by skaters and surfers who favoured the creativity of the shred (which only a board can provide) over pure alpine speed. And no matter the sport, culture, era or destination, all adventure athletes shared (and continue to share) that singular primordial lust to test oneself against nature and chase the tang of adrenaline itself.

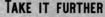

TAKE IT FURTHER
Push your limits:
Get out of your comfort zone, p180
Be empowered:
Face your fears, p266

SPEND A NIGHT IN THE JUNGLE

TAKE THE SLOW ROAD

VISIT DARK PLACES

UNLEASH YOUR CREATIVITY

SUPPORT A SOC

AY IT FORWARD

BE A KID AGAIN

GET LOST IN A CROWD

BE AWESTRUCK BY NATURE

DO LOOK DOWN

AL ENTERPRISE

MEET THE PLANET'S GIANTS

3

Withdraw and rejuvenate. Taking time out to focus on yourself offers valuable respite for the body and a tonic for the soul.

You've been staring out at the crashing surf off the Northern Californian coast for ... you don't know how long. You're not wearing a watch. The only thought that crosses your mind is making it back to the yoga studio in time for the sunset vinyasa flow. But you're also coming to a slow realisation: you feel at peace. Your mind has stopped its endless churning of worries and has become still.

There are as many reasons for going on retreat as there are retreaters. For some, those few days are a chance to focus on an aspect of physical health, to learn to break bad habits with professional support, or embark on a detox somewhere stunning. For others, it's an opportunity to awaken creativity: to hole up with like-minded people and workshop, brainstorm and nurture your imagination. For many more still, retreats can be some of the only time in the year to dedicate to a spiritual self.

From creative writing to qigong and chakra meditation to ayurveda, the variety of ways to take time out for a bit of self-care are vast. And you don't have to go far – there may be a gorgeous yoga studio in the Cook Islands, but there will also likely be a great mindfulness centre in your hometown. However and wherever you want to do it, regular retreating is said to deepen self-awareness, increase resilience and improve relationships. After all, when else do you get to spend so much quality time with yourself?

RETREAT

TAKE IT FURTHER

Savour an indulgence:
Upgrade, p168
Take your meditation to the next level:
Experience a week of silence, p258

Left: Looking out over Lake Wakatipu from Aro Hā yoga studio, New Zealand

PINK SPIRIT
Jordan
Sandra Jelly of Pink Spirit leads mindfulness retreats with the assistance of a troupe of Arabian horses. The otherworldly terrain of the southern Jordan desert doesn't hurt either. Retreat packages include sunrise yoga on the sandstone rocks of Little Petra, guided hiking, stargazing around a campfire and 'soul sessions' in the company of horses. **HOW** Yoga and horse therapy retreats include airport transfers and meals. www.pinkspiritjordan.com.

ARO HĀ
New Zealand
Aro Hā's philosophy is less about pampering and more about jump-starting a healthy lifestyle. Beginning with yoga, a typical day is packed with activities including hiking, strength training, mindfulness meditation and nutrition classes. The property's location overlooking Lake Wakatipu on New Zealand's south island adds to the improving vibe. **HOW** Aro Hā is a 45-minute drive from Queenstown. Week-long wellness and adventure packages available. www.aro-ha.com.

HOSHINOYA
Japan
Check in to this typical riverside *ryokan* (traditional inn) to try your hand at the contemplative practice of *kado* – the Japanese art of flower arrangement. The inn is located in a tranquil spot surrounded by gushing water and cypress trees and is accessed by boat. Lessons are given by a *kado* master and early-morning meditation is also on offer at a nearby temple. **HOW** Hoskinoya is located in **Arashiywama district of Kyoto.** www.hoshinoyakyoto.jp

TRAVEL WITH STRANGERS

Finally taking that long-awaited solo trip, but don't want it to mean a week, or even a year, spent in isolation? Want to be open to every experience and connection while travelling? Making a new buddy on the road is just what the doctor ordered.

Travel is about seeing new landscapes and eating delicious foods. But it isn't just about that. Getting out of the familiar routine of day-to-day life is also an opportunity to meet new people. It's a chance to break out of your shell, try out some rusty language skills, strike up conversations and forge new connections that can last well beyond your trip. If you're travelling solo, it's the perfect opportunity to join forces with someone in the same boat and share the experience; if sightseeing with a companion already, it's a way to get a better deal on that rental car for a planned day trip (split the cost four ways instead of two), or a chance to double-date with a couple from a different hemisphere. Authentic human connection adds richness and spice to any life experience, but be lucky enough to develop a friendship on the road and you'll be rewarded with laughter, support and honesty. too. What's more, you can gain self-awareness; the friends we keep are a reflection of who we are. Making travel buddies is surprisingly easy, it just takes a willingness to be a little social. Get comfortable with that, and it will be easier to practise back at home, too. Suddenly the world doesn't seem so big after all.

TAKE A WALKING TOUR

Cartagena de Indias, Colombia
Unesco World Heritage site Cartagena, with its walkable historic centre, is an ideal place to take a tour by foot and meet other travellers while doing so.
HOW Free Walking Tour Cartagena (www.freewalkingtourcartagena.co) runs daily tours at 9.30am and 4.30pm; meet at the clock tower plaza and tip your guide.

WORK REMOTELY

Cape Town, South Africa
Cape Town, with stunning Table Mountain set against the meeting point of the Atlantic and Indian oceans, is a regular stopping point on Remote Year itineraries, or simply a great base for anyone looking to take their remote-approved job to a new locale.
HOW Remote Year (www.remoteyear. com) programmes hook up communities of strangers to travel and work together.

Below (left and right): Making friends in the walled city of Cartagena de Indias, Colombia

MEET UP IN A FRIENDLY CITY

Gothenburg, Sweden
Gothenburg was ranked number one in the world for social milieu and values by Hostelworld, meaning residents are friendly, open and trusting. The port has Scandinavia's largest amusement park, perfect for visiting with new friends you met at the weekly Friday 'After Work' parties in bars across the city.
HOW For an upscale spot, try Yaki-Da (www.facebook.com/yakidagoteborg), or search Meetup (www.meetup.com).

JOIN A GROUP TOUR

South Asia
Tour company GAdventures is an option for people who want to meet new friends but don't want to travel completely alone. It runs itineraries aimed at 18- to 30-somethings.
HOW Try a trek through Indonesia or explore south India and Sri Lanka. See www. gadventures.com.

© ANDRESR | GETTY IMAGES, © MONICA HERNANDEZ AHMAN | SHUTTERSTOCK

From Peru to Prague with Wendulka

"The airfare to Lima was too tempting to pass up, and after booking my ticket I went straight to the Thorn Tree travel forum for itinerary-planning advice. That's where I found a post by Wendulka, almost exactly my age and seeking a companion in Peru for the same weeks I would be there. How can you argue against a coincidence like that?

I sent her a message and we agreed to meet in a Lima hostel the night we both arrived before setting off for the sand dunes of Huacachina together. We had our share of misadventures on the trip – a chipped tooth from a *granadilla* pip on one of our many overnight bus journeys led to my travel companion getting major dental surgery in Aguas Calientes just below Machu Picchu – but if anything our travel partnership only became stronger. Every day of the trip together went as smoothly as the one before, and later that year I visited my new pal in her home city of Prague.

I would always use caution when meeting people I've connected with online, and always arrange a first meeting in a public place such as a hostel. However, I've also discovered that there is something in the old saying that a stranger is just a friend you haven't met yet."

Nora Rawn, travel editor

TAKE IT FURTHER
Relish the disorientation of travel:
Experience culture shock, p172
Trust your own choices:
Take a big trip alone, p240

SUPPORT A SOCIAL ENTERPRISE

Some travel to seek natural beauty, some authentic flavour at meal times and a boozy nightcap with friends. Their goal is to chill. Which is fine. But there is a deeper, alternative way to wander.

When you actively look to see the world, in all its beauty and hardship, from someone else's perspective, travel becomes something much more than a holiday: it becomes an education. Social enterprise tourism offers the triple whammy of enabling travellers to have fun, learn about a culture and also benefit the community they are visiting. They are generally organisations created by and for the people who live in a given community or area. For travellers this could mean hiking the Inca Trail with indigenous Peruvian guides, taking music lessons with local musicians in Southern India, or purchasing handicrafts direct from the artisans themselves. Supporting a social enterprise involves travelling with an eye on a bigger prize than your own satisfaction and hip pocket and it will follow that you become more conscious of your impact whenever and wherever you roam. What's more, the experiences will almost certainly be accompanied by music, dance, great food and mutual understanding.

DESTINATION KARAKOL

Kyrgyzstan

Enjoy a food crawl in Kyrgyzstan's most flavourful city, led by locals who have lived and eaten there for years, if not all their lives. The 2.5-hour tour includes six tastings, each at a different restaurant. Your dollars go towards funding valuable projects, such as a public wi-fi network or a river park. It's best done between June and October.
HOW Most people fly into Bishkek and then transfer by ground transport to Karakol. See destinationkarakol.com.

MAJI MOTO MAASAI CULTURAL CAMP

Kenya

Posing with Maasai warriors in full regalia might make for a good photo, but most are snapped with scant cultural exchange. This Maasai owned and operated camp, based just outside the Masai Mara, 118 miles (190km) from Nairobi, supports vulnerable girls and women. Your dollars pay for health, education and conservation efforts that honour local practices. Some residents serve as knowledgeable wildlife guides.
HOW You can stay at the camp, where you'll experience traditional music, food and dance. Packages include transportation to/from Nairobi. See www.majimotomaasaicamp.com.

BLUE VENTURES

Belize

At the largest barrier reef in the western hemisphere, learn to identify endemic species, cull invasive lionfish, take reef samples and sharpen your dive skills while living in a traditional fishing village homestay. Blue Ventures helps the community manage and revitalise their fishery by providing accurate data, all of it collected by volunteers.
HOW Fly into Belize City and hop on a water taxi to Sarteneja where the journey begins. Prices vary according to your diving qualification and length of stay in Belize. See www. blueventures.org/volunteer/belize.

What is a social enterprise?

Social enterprises are organisations with the primary aim of making profit in order to fund social programmes. Simply put, they are businesses with a mission to help, and often they're locally owned. A social enterprise could be a local food crawl that raises money to engineer internet access, or a safari camp in East Africa that also helps educate and provide jobs for abused women and girls. It may have an environmental ethos. Whatever its vision, it's still a business designed to provide a specific service, which makes the relationship clean and, hopefully, sustainable for years to come. Sometimes doing a bit of lasting good is as easy as being a good customer. Check out the international volunteering and social enterprise database at www.grassrootsvolunteering. org to find recommended operators, vetted by your fellow globetrotters scattered all over the world.

TAKE IT FURTHER

Help the water world:
Become an ocean defender, p160
Get involved when it is most needed:
Help a community rebuild, p284

Left: Maji Moto Maasai Cultural Camp, near Narok, Kenya

© 2019 MICHAEL BENANAV | IMAGE COURTESY OF MAJI MOTO MAASAI CULTURAL CAMP

TAKE IT FURTHER

Have an immersive wildlife
experience:
Hide out to watch birds, p78
Observe the fragility
of existence:
**Respect the circle of life,
p288**

Left: A swarm of monarch butterflies during the migration in Mexico

It's a late-October morning in the highlands of Mexico's Michoacán state. The air is chilly, with a tinge of wood smoke from the houses down in the valley. You're lying in the dry grass, looking up at the rangy oyamel firs. The trunks of the trees, usually brown, are instead a rusty orange. Same for the needles, typically a deep piney green. Suddenly, the trees begin to ripple, as if covered in flames. A blanket of orange rises, undulating, into the sky. Then it settles towards the ground and you're surrounded by millions upon millions of monarch butterflies.

This is Mexico's Monarch Butterfly Biosphere Reserve, the wintering habitat for North America's monarchs, who flutter down from the northern USA each year. They cling to the sacred oyamel firs at night, then, warmed by the morning sun, descend to the ground to stay cool during the day.

A single butterfly is stunning, a work of art in miniature. But so many at once is awe-inspiring in an entirely different way. Such is the power of the swarm.

Swarms, flocks or massive herds of animals make for some of the earth's coolest travel sights. Watching creatures move in sync with each other and with invisible natural forces can be anything from an endorphin rush to a great photo opportunity to a full-on spiritual experience. It shows us just how small we are compared with the power of nature. Bring your camera and your willingness to be wonderstruck.

HOW The reserve is located about 62 miles (100km) northwest of Mexico City.

WITNESS A SWARM

Animals on their own are amazing. But get, say, 400,000 or four million of them together? That's something worth travelling the globe to see.

CAPELIN ROLL
Newfoundland and Labrador, Canada
Sometime in June or July, tens of thousands of tiny silver fish called capelin arrive en masse at the pebbled beaches outside St John's. Locals gather to watch them turn the seas to glitter, then scoop them into buckets to eat fresh. Humpback whales often follow, so bring binoculars.
HOW Middle Cove Beach in the provincial capital of St John's is usually a hotspot. Visit www.ecapelin.ca for information about the capelins' arrival.

BAT EXODUS
Malaysian Borneo
Those with chiroptophobia (fear of bats), stay away! Each dusk in Gunung Mulu National Park, millions of the flying mammals make their exodus from Deer Cave, swirling into the darkening sky like a strange tornado. Visit the yawning limestone cave itself first, home to imposing stalactites in addition to the 12 bat species.
HOW The nearest airport is in Miri. Tours of the cave run twice daily; book via the website: www.mulunationalpark.com.

WILDEBEEST MIGRATION
Kenya and Tanzania
Feel the thunder roll through your body as two million wildebeest, zebra and gazelle stampede across the Serengeti and the Masai Mara National Reserve each July and August in search of grazing land. Watch them from your 4WD, experiencing the thrill when they escape the clutches of a lion or Nile crocodile.
HOW Longstanding Arusha-based Roy Safaris (www.roysafaris.com) offers safaris lasting roughly two weeks.

CRAB MIGRATION
Christmas Island
In November or December every year, 100 million flaming red crabs march from their forest homes to their beach breeding grounds, turning the entire island scarlet. Traffic and human activity are brought to a halt as the crabs scurry across roads and bridges; there are even specially built crab bridges and tunnels to aid their movement. It's freaky and it's awesome.
HOW Fly here from Perth, Australia, or Jakarta, Indonesia. For predicted migrations and viewings, see www.christmas.net.au.

BE A KID AGAIN

Take a lesson from youth and tap into a child-like sense of wonder: be present, be silly and find adventure in the little things. Everyone's got a kid inside, and with a bit of encouragement they can ignite a sense of pure, unfiltered joy.

TAKE IT FURTHER

Be there for lift off:
See a rocket launch, p176
Have a thrilling animal encounter:
Look into the eyes of a predator, p202

MUSEUM OF SOVIET ARCADE GAMES

Moscow, Russia

Run riot with dozens of mostly functional Soviet arcade machines. Visitors get a paper bag full of 15-kopek Soviet coins, which fire up the recreational dinosaurs that would look at home in the oldest episodes of *Star Trek*.

HOW The museum is a three-minute walk from Kuznetsky Most metro station, open 11am–9pm daily. www.15kop.ru.

DARJEELING HIMALAYAN RAILWAY

India

This pot-sized antique steam train made its first journey along its precipice-topping tracks in 1881 and is one of the few hill railways still operating in India.

HOW The 'joy rides' from Darjeeling to Ghum and back take two hours. Tickets can be booked online at www.irctc.co.in or at the station in Darjeeling.

OLYMPIC BOBSLEIGH AND TOBOGGANING

Lillehammer, Norway

Bombing down a bobsleigh track, reaching speeds of 75mph (120km/h) and facing forces of 5G really looks like something you should leave to the experts. But in Lillehammer – Norway's oldest winter resort – they'll let almost anyone have a try.

HOW Lillehammer is 110 miles (180km) north of Oslo. The bobsleigh track is at Hunderfossen, 9 miles (15km) from town.

RICHARDSON ADVENTURE FARM CORN MAZE

Illinois, USA

A team of graphic artists redesign the maze on this farm to fanciful new heights of whimsy every year. Come September, it's all about the unbridled joy of running through the world's largest corn maze.

HOW The farm is 63 miles (102km) north of Chicago; open Sep and Oct only; www.richardsonadventurefarm.com.

Left: Tobogganing in Lillehammer, Norway

FOLLOW YOUR PASSION AROUND THE WORLD

Matt Phillips spent his youth chasing tennis balls while playing on the junior circuit in Canada, but in 2018 he fulfilled a long-held dream and chased his favourite tennis stars around the globe, taking in all four Slams in a single season.

Travel Goals

'*You* cannot be serious!'
The words of John McEnroe rattled out of the old, wood-lined Sony TV. It was 1981 and my eight-year-old self was sat in front of *Breakfast at Wimbledon* – and I was spellbound. My passion up until then had been hockey (I'm Canadian, after all) but all that changed with Borg vs McEnroe. Fast-forward a few years and my hockey stick had been replaced with a tennis racquet.

When visiting my sister in London in 2000, she surprised me with Wimbledon tickets. We cycled there and reminisced about all those days watching *Breakfast at Wimbledon* in our pyjamas. As I stepped on to Centre Court I was eight again, but now I was sitting inside that TV. Since moving to London, attending Wimbledon has become an annual affair, but during the other three Slams – Australian, French and US – it's 1981 again, just with a better TV. Longing to feel the energy inside those stadiums, the boisterous cheering in Melbourne, the slow clapping in Paris and the raucous night matches in New York, I set a goal in 2017 to visit all four the following year.

The Australian Open was everything I'd expected and more, with an incredible party-like atmosphere pervading the sun-drenched grounds during the day and lighting it up at night. I watched dozens of matches over numerous sessions, one in the famous 42°C (108°F) Melbourne heat. The chilly night matches were theatrical affairs, with laser-like light shows and music between games. It was all very surreal; at one point I even found myself sitting beside Rafael Nadal's father during one of his matches. When my modern tennis hero Roger Federer made the final, I took a deep breath and emptied my wallet for a ticket. Watching him win his 20th Slam in five dramatic sets will stick with me forever.

Hitting Paris for the French Open was a very different affair. Rather than throwing myself into the tournament as in Melbourne, this was to be a weekend taster. My girlfriend and I dovetailed a couple of matches at Roland-Garros with some sightseeing. The tennis on the red-clay courts was intoxicating, while the atmosphere in the stands was all rather regal.

Wimbledon started as it always does, with a pre-dawn arrival on the second Tuesday (tent and sleeping bag in hand) to camp for Centre Court tickets

on Wednesday. After a day of Frisbee and lazing in Wimbledon Park, we found ourselves courtside for what were truly epic quarter-finals: Nadal vs del Potro and Djokovic vs Nishikori.

New York was my last Grand Slam stop. The energy in the grounds and stadium was electric, and I thought I'd hit the jackpot when Federer was slotted to play in our night session. But in life, like sport, there are winners and there are losers. That evening, Federer and I were the latter. It was a gutting way to end my Grand Slam tour, but it also emphasised just how fortunate I was to see him lift the trophy in Melbourne.

Looking back at the year, I'm grateful as it has bolstered my love of the game. The year's top moment? Sitting in the front row of the Rod Laver Arena and watching the still-masterful John McEnroe tear up opponents during a doubles match late one evening.

Yes, I'm serious.

HOW The Aussie Open takes place over the last two weeks of January; Roland-Garros from late May to early June; Wimbledon in early July; and the US Open from late August to early September.

Above (left and right): The blue court of the Australian Open; racing through the Monte Carlo Grand Prix.
Previous page: Roger Federer takes aim during the Australian Open

"As I stepped on to Centre Court I was eight again, but now I was sitting inside that TV."

MOTOR RACING'S TRIPLE CROWN

The Monaco Grand Prix, Indianapolis 500 and 24 Hours of Le Mans together make up motor racing's unofficial Triple Crown. Each is different in spirit, locale and class of car, but all are entrancing to watch. Only one driver has won the Triple Crown (Englishman Graham Hill), but attending all three is a far easier proposition.

HOW Check the websites for race details: Monaco GP (www.monaco-grand-prix.com), Indy 500 (www.indianapolismotorspeedway.com) and 24 Hours of Le Mans (www.lemans.org).

THE WHISKY TRAIL

It's been around for centuries, but whisky is currently experiencing a global boom. Whether in Scotland, Ireland, the USA, Canada or Japan, there's now an intoxicating variety of options to taste. If you have a passion for this elixir, then why not sniff and sip your way around the globe to enjoy it.

HOW A great start would be in whisky's birthplace, Scotland, where there are more than 120 active distilleries (20 more will open by 2020). Check out Scotland's Whisky Map at www.visitscotland.com.

FILM FESTIVALS

Festivals to celebrate film mix art and creativity with varying doses of celebrity. Travelling to the big hitters of Venice, Cannes, Berlin and Toronto will put you face to face with the big stars, while smaller-scale treats such as Sicily's Taormina Film Festival, which holds evening screenings in an ancient Greek outdoor theatre, are more intimate affairs.

HOW The New York Film Academy has a comprehensive list of the world's festivals, with links to each. www.nyfa.edu/student-resources/film-festivals.

LEARN FROM IND

Signing up for an indigenous tourism experience won't just open your mind, it will also empower your hosts to preserve their rich cultural heritage.

GENOUS CULTURES

It's one thing to marvel at cultural relics through glass cabinets in a museum, quite another to meet the living descendants of the world's early cultures, many of whom continue to practise their ancient traditions to this day. Responsibly managed indigenous tourism provides an incredible platform to interact with First Nations peoples in a meaningful way, leaving you with a fascinating insight into – and no doubt a greater respect for – the cultural heritage of your destination. In turn, your participation encourages local people to keep their culture alive and pass it on to future generations.

Of the many forms of indigenous tourism, community tourism is among the most popular. This typically involves visiting an indigenous village to learn about and celebrate cultural traditions – from how to search for bush foods to observing how traditional artwork is made. Employing the services of indigenous guides and shopping at indigenous-run art centres are also great ways to gain a deeper understanding of, and give back to, the indigenous peoples of your destination.

It's essential to do your research to ensure any experience you are considering is truly respectful of the local community and empowers your hosts both economically and culturally – if you're booking through an operator, be sure to ask what percentage of the fee the community receives. When visiting any indigenous community, research the most respectful way to dress and interact with locals, and always ask permission before taking photographs.

TAKE IT FURTHER

Commune with your ancestors:
Seek out the relics of an ancient civilisation, p178
Gain a mutual understanding:
Master a foreign tongue, p260

Left: San woman, southern Namibia

LIVING MUSEUMS
Namibia
Spending time with the San people, the longest-standing inhabitants of southern Africa, is one of the most rewarding parts of visiting Namibia or Botswana. Run by the Living Culture Foundation of Namibia (www.lcfn.info), Namibia's five Living Museums offer visitors a chance to see how the small-statured San people, who speak with a distinct clicking noise, have survived off the land for centuries; it also provides meaningful employment to local communities.
HOW Intrepid Travel (www.intrepidtravel.com) includes a visit to a Living Museum in several of its Africa overland itineraries.

ULURU
Northern Territory, Australia
Join a member of the Uluru family on a tour of their traditional homelands just outside Uluru-Kata Tjuta National Park. After searching for bush tucker, you'll learn about the instrumental role the family has played in the battle for Aboriginal land rights, and watch the sun set over Australia's most famous rock from a stunning viewpoint far from the tourist crowds.
HOW Seit Outback Australia's Patji Tour (www.seitoutbackaustralia.com.au) lasts seven hours.

ALASKA NATIVE HERITAGE CENTER
Anchorage, USA
Many of Alaska's remote Native communities lack the infrastructure to host tourists. This makes a visit to the Alaska Native Heritage Center in Anchorage that much more special. At this wonderful facility, visitors can stroll through six authentic life-sized Alaska Native dwellings and meet representatives from each culture at each site. You can even opt to purchase art from Native vendors.
HOW The centre (www.alaskanative.net) is 7 miles (11km) from Anchorage. Summer shuttles run from the Visit Anchorage office on 524 W 4th Ave, the Anchorage Museum and several hotels.

BE AWESTRUCK BY NATURE

Awe is a complex and little-understood emotion, but we know it when we feel it. New research shows it may actually be good for us. One of its finest sources? Nature, in all its vast and unknowable glory.

*E*ncounters with the awe-inspiring are not just cool, they can actually make us better people. 'Fleeting and rare, experiences of awe can change the course of a life in profound and permanent ways,' write psychologists Dacher Keltner and Jonathan Haidt, in a seminal 2003 paper on the emotion. Awe, Keltner and Haidt conclude, has to do with a sense of vastness, of something far bigger than yourself. It also involves a feeling that you're encountering something new and outside your frame of reference. Having such an experience can instantly transform your life goals and values, potentially re-orienting you towards a better and less self-centred way of being.

Nature is one of the most common sources of awe, just as it was at the dawn of humanity. So much of it is bigger than us: thunderous waterfalls, whales the size of city buses, deserts of strange knobby pinnacles stretching as far as your binoculars can see. And so much of it is foreign to our human minds: the way thousands of starlings know to take to the sky all at once, salt flats like Martian landscapes, the unreadable gleam in the giant eye of an octopus as she whooshes across the sea floor.

SALAR DE UYUNI

Bolivia
At nearly 12,000ft (3660m) above sea level, the world's largest salt flat is a harsh and pitiless landscape. Its 4633 sq miles (12,000 sq km) of bone-white desert will blind you by day and freeze you by night. Come for the sense of being at the far edge of a mystical land. Jump off from the town of Uyuni.
HOW Operators offer tours of the salar. Try Cordillera Traveller (cordilleratraveller.com).

FRANZ JOSEF GLACIER

New Zealand
It's shrinking, of course, which is all the more reason to see the South Island's most famous glacier. Strap on a pair of crampons for a tour over the diamond-bright ice, climbing up ladders and through blue ice caves.
HOW Franz Josef Glacier Guides (www.franz josefglacier.com) offers half-day heli-hikes.

NAMIBRAND NATURE RESERVE

Namibia
This wild stretch of savannah and sand dunes is awe-striking in the day, when oryx and springbok thunder across the golden landscape. But come nightfall the park turns even more magical, with one of the darkest, most star-spangled skies on earth. Marvel at the infinite cosmos with a talk from a resident astronomer at the Sossusvlei Desert Lodge.
HOW The private reserve (www.namibrand.com/tourism.html) is a five-hour drive from Windhoek. Visit on a guided safari.

TONGUE OF THE OCEAN

Andros Island, Bahamas
Off the coast of Andros the shallow sea suddenly plunges into a 6600ft (2010m) trench called the Tongue of the Ocean. On a wall dive you'll sink from the aqua waters towards unimaginable depths, watching large predatory fish play along the cliff as you descend.
HOW Small Hope Bay (www.smallhope.com) is a favourite resort of divers, with courses offered.

Left: Oryx in the NamibRand Nature Reserve, Namibia

© MARTIN HARVEY | ALAMY STOCK PHOTO

Grasshoppers & awe on the great plains

"We were in western Oklahoma when the grasshopper storm hit. At first I thought it was a fat raindrop hitting my head as I stood at a decaying service station off I-40. Then I felt another, and another. Too big to be rain. I jumped back in the car and they were upon us, thousands of insects smacking against my windshield.

My friend and I had decided a drive from North Carolina to San Francisco was in order. We'd figured the Great Plains would be a yawn. But as soon as we crossed from the greenness of Arkansas and the land began to open, I became uneasy. I'd always lived among trees and hills, and suddenly I felt tiny and exposed. There was too much sky, and then, alarmingly, too many grasshoppers. But as we passed into the Texas Panhandle, my anxiety faded. The sun was setting, turning the plains an otherworldly orange. Storm clouds purpled the horizon, distant lightning strikes cracking. The sky no longer felt too big; it felt boundless, full of magic and possibility. I no longer felt fear. I felt awe."
Emily Matchar

TAKE IT FURTHER

Get active and educated:
Preserve the planet, p228
Appreciate Mother Earth in all her magic and majesty:
Witness a miracle of nature, p282

Romance is thrilling enough at home, but add a foreign locale as the backdrop and the scene is set for an unforgettable experience, whether your new beau is from the same city or from all the way across the globe.

TAKE IT FURTHER

Travel proud and party:
Embrace your sexuality, p166
Find your people and stay a while:
Experience community living, p238

An openness to travel often means a greater willingness to experiment and take risks – maybe that's why a great trip pairs so well with a great fling. Take the usual precautions and pay attention to your instincts, but don't close yourself off to the opportunity to connect romantically on the road. When else is it possible to ride tandem on a motorcycle as the sun sets while sharing a set of earbuds? Whether it's through using a dating app abroad or simply sharing a seat on the bus or crossing paths several times in the same town, the opportunities for chance meetings are many. Strike up a conversation and see where things go – maybe there's a shared destination to join forces in visiting, or else it's the perfect chance to swap language lessons while canoodling. Handsome strangers need someone to keep them company just like anyone else! When parting ways, don't forget to share your contact information to potentially reunite later on. Some sparks are meant to fade, while others blaze into something more. Either way, the memories will be the best souvenir anyone could ever ask for.

And what if you're already coupled? Studies show that novelty, new surroundings and especially adrenaline add zest to a relationship, so use your trip to try that adventure sport you've always been meaning to give a go. Hearts beating together as you whizz down a zipline or windsail into the sunset will add a frisson to any relationship.

HAVE A FLING

MARDI GRAS
Sydney, Australia
If having a fling is on your agenda, why not head straight for Sydney's annual Gay and Lesbian Mardi Gras? At Australia's biggest LGBTQ event, hedonism and celebration are the top two items on the agenda and there's no shortage of ways to get involved. It's held annually every February and March.
HOW A two-week festival leads up to the parade; register in advance to walk in a group or watch the floats from Oxford or Flinders St.

BEACHES OF THE YUCATAN
Mexico
The Yucatan Peninsula offers everything a romance-inclined traveller could want: beautiful beaches, stunning Mayan ruins, endless cenotes and delectable fresh tropical fruit for every morning's breakfast. Convenient access to the airport equals plenty of travellers in the region, and there are lots of hammocks to get cosy in.
HOW Stay at Poc-Na Hostel (www.pocna.com) on Isla Mujeres for a laid-back island vibe, or enjoy the poolside party at Hostel Quetzal (www.hostelquetzal.com) in downtown Cancún.

GULET CRUISE
Turkey
Confined spaces breed intimacy. If you want to up the ante on a new connection, take the paragliding excursion at Ölüdeniz while on a gulet cruise along Turkey's beautiful turquoise coast. Nearby Mt Babadag is one of the world's best spots for this adrenaline-pumping activity, a scientifically proven way to increase your pair-bonding with that attractive stranger.
HOW Cruises depart Kas, Fethiye or Olympos and spend two–three days cruising the islands of the Lycian Way. Try Alaturka Cruises (www.alaturkacruises.com).

AKIHABARA MAID CAFES
Tokyo, Japan
Want to make a connection but prefer to keep it strictly platonic? It won't develop into a more meaningful relationship ('maids' don't even use their real names or reveal their hometowns), but the strange, admittedly paid, intimacy of one of Tokyo's maid cafes has its own thrill, plus it comes with an adorable drink.
HOW Try @Home Cafe, 4th–7th fl, 1-11-4 Soto-Kanda, Chiyoda-ku, Tokyo; open until 10pm daily.

PAY IT FORWARD

POTATO PRESERVATION

Pisac, Peru

A tour company that builds itineraries around social enterprises will help you to support local initiatives. Travellers to Parque de la Papa (Potato Park) in Pisac learn about rural life in Peru, while the Andean community run a seed conservation programme, conserving local potato varieties.

HOW G Adventures (www.gadventures.com), teamed with The Planeterra Foundation, provide socially responsible tours.

WOMEN'S EMPOWERMENT

Kerala, India

Women in India face many social inequalities, from gender-specific abortions to forced marriage and denial of education – 68% of India's illiterate adults are women. Teaching skills such as income generation is a vital step towards equality.

HOW GVI (www.gviusa.com) offers volunteer opportunities to promote women's empowerment, including a two-week project in Kerala.

ELEPHANT REHAB

Sri Lanka

While the wild elephant population of Sri Lanka has dwindled, domesticated elephants are still used for logging, construction and temple work. Support the welfare organisations that care for them, by volunteering or having a non-exploitative tourist experience.

HOW The Millennium Elephant Foundation (www.millenniumelephantfoundation.com) offers schemes lasting from one day to six weeks.

REFUGEE CRISIS CALL

Greece

The tiny island of Lesbos in Greece is bearing the brunt of Europe's refugee crisis, with almost 400,000 arrivals (to an island of 90,000 people) since 2015. Groups of caring citizens are helping in the refugee camps, and they need volunteers.

HOW Find volunteer opportunities on Lesbos at One Happy Family (www.ohf-lesvos.org) and Sea of Solidarity (www.seaofsolidarity.org); many others organise volunteers around Europe.

BUY LOCAL

Siem Reap, Cambodia

Whenever you can, shop for souvenirs at a fair-trade cooperative so that the majority of the money that you pay goes to the maker. In well-organised operations even the re-seller will often be a local. It really does pay to do your research before you go.

HOW Siem Reap's Artisans Angkor (www.artisansdangkor.com) trains young craftspeople to keep traditional Khmer skills alive.

TAKE IT FURTHER

Protect the vulnerable::
**Help save an
endangered species, p232**
Volunteer long-term:
**Give a year of your
life for others, p280**

We're the lucky ones, the ones who travel for pleasure. It's a privilege to discover the world, so when we get there it's our duty to do what we can to improve the lives of those – people and animals – who don't have the same good fortune as we do.

*Above: Elephants
in Sri Lanka*

SAVE THE KAKAPO

New Zealand

Pity the kapako. They once teemed in New Zealand, but when Europeans and their boats full of dogs, cats and rats arrived, the birds' flightlessness (and 9lb (4kg) heft) made them easy prey. Today, there are fewer than 200 of these chunky parrots left. Volunteers can support efforts to keep them breeding.
HOW Opportunities are listed at www.doc. govt.nz/our-work/ kakapo-recovery; breeding season is January–May.

MUMBAI SLUM LIFE

India

Are slum tours purely exploitative or do they offer opportunities to slum-dwellers and perspective to tourists? Tours of Dharavi slum, one of the world's largest, showcase the community's diversity and spirit, and 80% of profits go back to the community.
HOW Read up on the pros and cons at www.slumtourism. net; a Dharavi tour can be booked via Reality Tours & Travel (www. realitytoursand travel.com).

PRECIOUS CARGO

Various

Making a difference when you travel can be so easy that you barely notice it. Alongside your travel essentials, pack coloured pencils for schoolkids, bandages and surgical gloves for medical centres, gardening tools for conservation programmes. These are small things, but life-changing for communities in need.
HOW Find out who needs what, where at Pack for a Purpose (www. packforapurpose.org).

SPREAD THE WORD

Luang Prabang, Laos

Laos is a deeply Buddhist country, and many men spend part of their life, usually between school and starting a career or getting married, living as a temple monk. Extra education during this time – for example, learning to speak English – can help their future employment and quality of life.
HOW Volunteer to teach English to monks in Luang Prabang, Laos' spiritual capital; GVI (www.gviusa.com) lists opportunities.

GLOBAL DOG RESCUE

Various

Hordes of street dogs roam the towns and villages of Nepal, Thailand, Guatemala and Mexico. Born strays or abandoned pets, they're neglected, hungry and often sick; in some cases they're a risk to local humans. Volunteer programmes provide community education, spaying and neutering clinics, and hugs and pats to mutts desperately in need of a little love.
HOW Dog rescue schemes are listed at www.volunteerhq.org.

In the depths of a forest that swarms with wildlife, your position in the food chain is all too clear. To be truly humbled by nature's might, stay overnight in the heart of the jungle.

SPEND A NIGHT

Even in broad daylight, the jungle can feel like an unfathomable place. But spending the night in the thick of a forest can awaken our most primal fears. As the forest darkens, the chorus of cicadas rises to a near-deafening buzz. A dull thud on the roof of your hut could be a tree branch, or a nocturnal animal landing from above. And, in habitats stalked by big cats, a rustling sound outside the door at midnight is better left uninvestigated.

Somewhere between tropical rain and unknown mammals scrambling across the rafters of your lodge, a humbling realisation dawns: human rules don't apply here. In Southeast Asia's jungles, where sun bears tear at tree bark and tapir mark their territory, humans seem like feeble specimens. Against the Amazon basin's 400-plus mammal species, we represent a mere speck within teeming, interlaced ecosystems.

As the hours pass during a night in the jungle, nature encroaches ever closer: ants march without rest along the window frame, or a pair of beady eyes outside catches the light. When a thunderstorm builds to its ear-splitting crescendo, it becomes clear you're at the mercy of nature's powers.

But gradually you grow accustomed to the sounds of the forest. Your eyes adjust to the low light and your senses feel keener. Amid the jungle's racket of hooting birds, croaking frogs and scuttling lizards, sleep can be elusive – but a feeling of awe endures long after dawn breaks.

IN THE JUNGLE

TAKE IT FURTHER

Challenge yourself:
Eat insects, p146
Get comfortable with the unknown:
Descend into the abyss, p206

SEPILOK JUNGLE RESORT
Malaysian Borneo
Some of the world's oldest rainforests cloak Borneo, the largest island in Asia. Clouded leopards pad through its jungles, hornbills soar above the canopy, and bats pour from caves at sunset. In northerly Sabah, Sepilok Jungle Resort (www.sepilokjungleresort.com) has private huts and dorm accommodation allowing travellers to doze to the sound of scampering geckos.
HOW Meet humankind's closest relatives just a five-minute walk from the resort; 80 orphaned primates swing freely through Sepilok Orangutan Rehabilitation Centre.

LAPA RIOS
Costa Rica
Scarlet macaws, sloths and giant anteaters are just a few species in Costa Rica's impressive jungle menagerie (in all, 5% of the planet's biodiversity). Ecolodge Lapa Rios (www.laparios.com) on the Osa Peninsula, which boasts the government's maximum 'five leaf' rating for sustainability, has bungalows tucked into a 1000-acre lowland rainforest reserve.
HOW Roads from San José to Lapa Rios are maintained, but the lodge can arrange air transport if you aren't keen on the seven-hour drive.

CRADLE MOUNTAIN WILDERNESS VILLAGE,
Australia
Cool-climate travellers can plunge themselves into rainforest wilderness in Australia's island state. Cradle Mountain Wilderness Village (www.cradlevillage.com.au) has private cabins hidden among northwestern Tasmania's moss-draped trees, where there's a good chance of playing hide-and-seek with pademelon. If you hear something big go bump in the night, it might be a wombat.
HOW There are numerous walks around the resort. To enter Cradle Mountain National Park, shuttle buses run from the park visitor centre to Dove Lake.

UNLEASH YOUR CREATIVITY

Just ask the world's greatest poets and painters: when you shake up your routine with travel, it invigorates the brain and opens the floodgates of creativity.

Artists have long sought the creative benefits of travel. Vincent van Gogh journeyed to the south of France to inspire his greatest swirls of colour. Paul Gauguin sailed for Tahiti to spark his sultry landscapes. Ernest Hemingway, F Scott Fitzgerald, Gertrude Stein and other writers of the Lost Generation used bohemian Paris to stoke their imaginations.

Turns out these folks were on to something. Scientists have established a connection between creativity and travel. It has to do with the brain's neural pathways, which are shaped by environment and habit. Travel changes your surroundings, and so changes the pathways. That is, foreign experiences – new sounds, smells, language, tastes and sights – trigger different synapses in the mind that rewire it and refresh it.

No wonder you feel the urge to write, sketch, dance, draw or otherwise experiment artistically on the road. It's only natural. Plus, you have more free time to explore your creative side and access to classes inspired by your location – a wood carving workshop in Indonesia, say, or a fiddle-playing lesson in Ireland. You just never know what sort of hidden talents will be set free when you globe trot.

TANGO

Buenos Aires, Argentina
Some people spend a lifetime trying to master the steamy strut, which has been described as making love in the vertical position. Tango is about capturing beauty and passion in movement, with each dancer bringing his or her interpretation to the parquet floor.
HOW Learn tango in its homeland at DNI Tango School (www.dni-tango.com) in Buenos Aires. Newbies can start with the 1.5-hour beginner's class and work up to advanced.

Left: Tango dancing in Argentina

PHOTOGRAPHY

Edinburgh, Scotland, UK
Photography is about seeing things in a new way, much like travel itself. Getting behind the lens and expressing your imagination through composition and lighting invigorates your creative side. There are loads of workshops to choose from that combine instruction with exotic locations.
HOW Take a three-hour lesson with Iconic Tours (www.iconictours.co.uk) in Edinburgh. A professional photographer shows you the city's main sights while also teaching you how to shoot them with style.

IMPROV

Chicago, USA
Learning improvisational comedy is a tried-and-true creativity booster. It pushes you to take a leap of faith, trust your instincts and embrace your mistakes – qualities that are valuable in any profession. Chicago, Los Angeles, New York, Toronto, London and Sydney are hubs for improv, but you can find classes all over the globe.
HOW Immerse yourself in the scene at iO Theater (www.ioimprov.com) in Chicago, the birthplace of improv; five-day intensive workshops happen every three months.

Stressed out? Be creative about it

The next time you're feeling overwhelmed, paint a picture. Or write a haiku, doodle in a notebook or make a collage. Studies show that engaging in a creative activity, whatever it may be, reduces stress. That's because it forces the mind to focus on the task at hand and puts you in a state of flow, where you're so absorbed in what you're doing that you can't think of anything else. It becomes a type of meditation. So don't worry if your watercolour of a tree looks like an abstract turtle, or if your crocheted mitten is missing the thumb. Sometimes, it's the process of creating that matters, not the end product.

TAKE IT FURTHER

Travel slow and with style:
Take a vintage transport method, p186
Set your body free:
Strip off, p248

When you travel by winding backroads, dawdling boats or other slow-going means of getting around, you forge a deeper connection with your destination and get to savour things – such as rolling vineyards, tiny villages, the World's Largest Ball of Twine – you'd miss if just jetting by.

Going fast often feels like the only way to go. You have one week for your trip, who knows if or when you'll be travelling again, so you'd better shift to high speed to see it all. Race along the highway, hop on a flight – do whatever is quickest to get from one place to the next.

Yet here's the thing: going slow adds so much more to the experience. You can fly between islands, but voyaging by water allows you to pause to drink at pirate bars and snorkel over shipwrecks. Leisurely train rides enable you to become immersed in the landscape and get a taste of local life at station stops.

Two-lane byways carry you to sweet homely eateries for your meal, rather than dining at soulless fast-food restaurants on the toll road. You can slow it down even more if you opt to travel by foot, by horseback, by canal boat, canoe or bicycle.

So stop rushing around. Slow travel – like the overarching slow movement – is about quality over quantity, about doing things as well as possible instead of as fast as possible. It's OK to take the slow boat to China or the world's slowest express train in Switzerland. The goal is to make the journey as meaningful as the destination.

TAKE THE S

TAKE A FELUCCA DOWN THE RIVER NILE

Egypt

Small Egyptian sailing boats called feluccas rely solely on the breeze and currents of the river. The traditional vessels are a magical way to soak up the Nile's beauty: the soft light of sunrise, fishermen casting their nets, lazy spots to go swimming.

HOW Dally on a three-day felucca trip from Aswan to Edfu. Captains typically hang out in Aswan's Nile-side restaurants.

RIDE THE TRAIN THROUGH HILL COUNTRY

Sri Lanka

The colonial-era train that chugs between Kandy and Ella is one of the world's most beautiful rides. Tea plantations, waterfalls and mist-wrapped emerald peaks drift by. At stations along the way, sellers jump on to hawk chilli fritters, fresh-cut mango with cinnamon and other treats.

HOW The trip takes six to seven hours; there are five trains daily. See www.railway.gov.lk.

WALK TRAILS IN LAS ALPUJARRAS

Spain

In the Sierra Nevada range, the region of Las Alpujarras mixes woodland, terraced farmland and Moorish-style white villages clinging to hillsides. User-friendly trails roam between the beguiling little towns, which all have guesthouses and restaurants, making multiday walks easy.

HOW Start in Pampaneira (near Granada) and set out on the iconic GR7 path. Nevadensis (www.nevadensis.com) has maps and arranges guided hikes.

TAKE IT FURTHER

Know the freedom of a self-propelled journey:
Travel by bike, p170
Slow down to nature's pace:
Live off the land, p272

Below: Traditional feluccas catching the last of the sun's rays on the river Nile

⌊OW ROAD

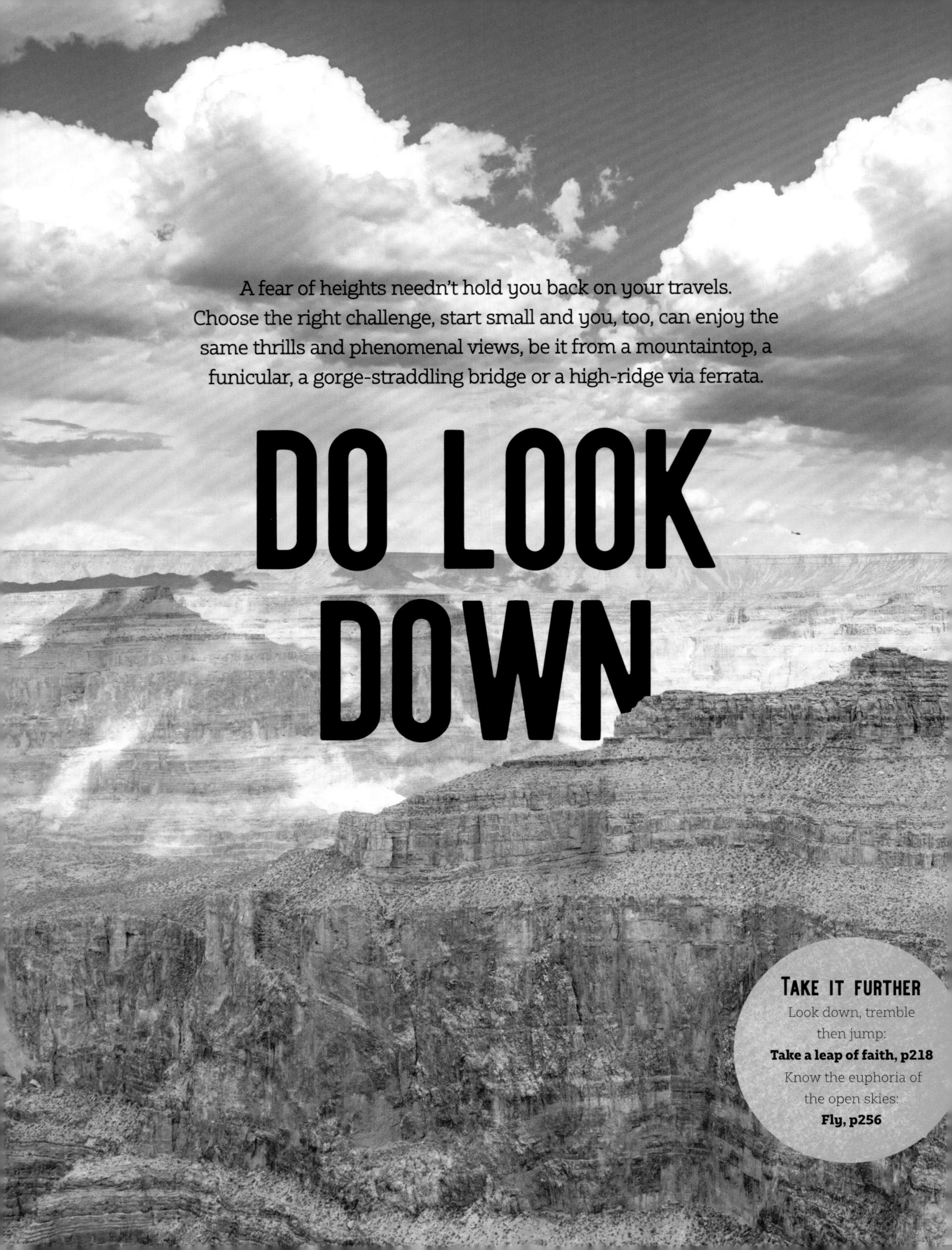

A fear of heights needn't hold you back on your travels. Choose the right challenge, start small and you, too, can enjoy the same thrills and phenomenal views, be it from a mountaintop, a funicular, a gorge-straddling bridge or a high-ridge via ferrata.

DO LOOK DOWN

TAKE IT FURTHER

Look down, tremble
then jump:
Take a leap of faith, p218

Know the euphoria of
the open skies:
Fly, p256

If standing on the edge of a sheer precipice and peering down into the void chills your blood, makes you break out into a cold sweat or erupt with expletives – you, dear friend, are not alone. For an acrophobic, that razor-sharp edge and gaping space below is not a bit of fun or an adrenaline rush, it's a mere footstep away from sure, sudden death. The trembling knees, the dizziness, the skip-a-beat heart, the relentless gut-churning, the sweaty palms – it's all very, very real and it's not going away until you make it back down in one piece.

But it's possible to get the 'height buzz' that others rave about without risking life and limb – start small, safe and with a good guide should be your mantra.

You might never make it to the top of Everest, but hey, with a rope and well-secured karabiner or an experienced guide, you could perhaps flirt with mountaineering on a via ferrata or canyon down a waterfall. If that sounds too intrepid, there are other (no less worthy) challenges to haul you that bit higher: from treetop walkways and suspension bridges spanning gorges to city skyscraper lookouts, cable car rides in the Alps, great glass elevators and high-level hikes that stick well away from the edge. Scenic flights are another way to get the views with all the brow-wiping phews.

Challenge yourself on your travels by beginning at the right level, build up confidence gradually, and you'll soon look down but never back.

VIA FERRATA
Mürren, Switzerland
Before ruling rock climbing out entirely, consider a via ferrata (fixed rope route) – giving you the intrepid buzz and sensational views without the risk, thanks to a clipped-on karabiner. Head to the via ferrata in Mürren in Switzerland's lovely Jungfrau region. From mid-June to October, the 1.4-mile (2.2km) trail offers head-spinning highs on cables, ladders and a suspension bridge with arresting views of the glacier-streaked Eiger North Face.
HOW Klettersteig (www. klettersteig-muerren.ch) can arrange it for you.

Left: Eagle Point, Grand Canyon

HOT-AIR BALLOONING
Serengeti, Tanzania
A qualified pilot guides the hot-air balloon ever higher and the drop barely registers as you focus on the landscape unfolding. Throw in a little wildlife on a ride over the Serengeti and you'll be so absorbed by down below you'll have forgotten any fear. Hop on at dawn for a glimpse of zebra, elephants, giraffes, lions and cheetahs. Flights are year-round but are most spectacular during the wildebeest migration (roughly March to June).
HOW It's not cheap but, hey, you're floating over the Serengeti! Find out more at www.balloonsafaris.com.

OBSERVATION DECK
Burj Khalifa, Dubai
Manmade structures (with solid ground beneath your feet) can be a non-nail-biting way to get high and confront your fear. High-rise observation decks poke above some of the world's metropolises and afford dizzying views of the cityscape through glass walls and, in some cases, floors. Why not go straight to the top of world's tallest building? Dubai's needle-thin Burj Khalifa (www. burjkhalifa.ae) pierces the sky at 2716ft (828m).
HOW Rise to Burj Khalifa's levels 124 and 125 on a self-guided visit. Book in advance online.

HIGH WALK
Great Canyon, USA
With something to cling firmly to, an above-the-treetops suspension bridge, a glass walkway or a hanging bridge can be fun ways to battle acrophobia without any danger of slipping. For a true sense of the monstrous scale of the Grand Canyon's West Rim, you can walk on glass 4000ft (1200m) above it on the Skywalk at Eagle Point. Rest assured, this glass-floored, horseshoe-shaped walkway can withstand the weight of 70 747 passenger jets.
HOW See www. grandcanyonwest.com for info about guided visits.

MEET THE PLANET'S GIANTS

Those who reside in the icy northern realm of the polar bear live in a permanent state of trepidation, and daily life is adapted to avoid chance meetings with the planet's largest land carnivore. Yet, when finally faced with a polar bear, **Sarah Barrell** felt not terror but pure soul-lifting awe.

A siren cuts through the frozen air. I look over at Doug as we press through the sideways sleet towards the Tundra Inn. He nods but doesn't pick up the pace. 'The siren signals voluntary curfew,' he says. 'Mainly so kids know it's time to come home.' The air-raid-style alarm also indicates the start of dawn-to-dusk patrols, during which men with rifles scour the streets for signs of ursine life encroaching on the remote Canadian town of Churchill.

Like all great apex predators, a polar bear's presence is felt even when it's not there. It's as if the creature still casts a shadow long after it's gone, humans gingerly stepping around the ghost of its shape, eyes always cast over shoulders; the ever-present promise of its appearance driving the dictates of day-to-day existence.

Tonight, boreal storms may have blown away imprints of the ice bears' hubcap-sized paws on Churchill's snowy shores, but hazard signs alert me to their preferred path between the boulders along the beach. The unlocked cars lining city streets point not, I learn, to a notable sense of community but also of the need for conveniently parked panic rooms to service pedestrians in the face of a surprise charge from the planet's largest land carnivore.

'Mostly they don't like town,' says Doug, one of Churchill's expert wildlife guides. 'It's noisy and smells funny.' Still, some do, notably unpredictable juveniles, and even though you can't see inside the 'polar bear jail', the former aircraft hanger where strays are held before being helicopter released, it's one of the first stops for the thousands of tourists who descend on Churchill during polar bear season. But a day spent bumping around Churchill's boggy bay in an all-terrain Tundra Buggy is the best way to track them, albeit slowly, painfully, patiently. My fingers freeze, poised on camera buttons, and my eyes stream while trained through binoculars, scanning the sleet-blasted horizon, eyelashes forming icicles in the Arctic gusts that lash the truck's open viewing platform.

Minutes, hours tick by. We become faceless shapes, retreated inside giant hooded jackets. And then a yelp goes up. Silence is golden when wildlife spotting but the thrill is too sharp. A sighting. It's distant – maybe 1300ft (400m) away – an average-sized female, thinks Doug. She's curled up, nose in paws, coal eyes closed against the wind. She raises her head intermittently,

quizzical like a puppy woken from a nap, licking at the life-sustaining nutrients in the kelp bed beneath, she yawns and replaces her giant head. I could sit for hours staring at her endearing boulder-like bulk but the radio buzzes. Another buggy has a sighting.

We find the beast quickly this time: a young male, pawing at the kelp with claws like garden rakes, its comically mobile but preternaturally powerful nose roving around for our scent. We get within 65ft (20m) – he's huge but not nearly close to the 2000lb (900kg) of a fully grown male, according to Doug. For 20 mesmerising minutes, we watch him sniff and scrape around the shoreline. And then suddenly he's moving towards us. Fast.

He's at the truck in seconds. I've no time to refocus my camera. He's too close. He rears up, paws on the truck's side, head straining over the edge less than a metre from my face. I squeak, a strangled scream, and instinctively step back. He fixes me with his eyes. My pulse hammers in every vein, but it's not fear; it's overwhelming awe, breath-taking joy, love even – an immensity of emotion directly proportional to his size. I feel as if for the first time, somehow, I've really been seen by another living thing. It's the most gentle, piercing and pure stare I've ever encountered.

Doug looks at me, as misty-eyed as I am. 'It doesn't matter how many times they look at you,' says Doug. 'It changes you forever.'

"My pulse hammers in every vein, but it's not fear; it's overwelming awe, breath-taking joy, love even"

Above (left and right):
Walking along the Hudson Bay; redwoods soar in Muir Woods, California.
Previous page: A polar bear in Churchill, Canada

SEE POLAR BEARS
Churchill, Canada
From September to November, after months without food, hundreds of polar bears amass on the shores of Churchill, a remote Canadian town on the Hudson Bay, waiting for sea ice to form so they can set off on winter hunts.
HOW Frontiers North offers buggy excursions into the surrounding Churchill Wildlife Management Area; you can stay overnight at its Tundra Buggy Lodge. Each November it hosts an epic 12-day expedition into the nearby Wapusk National Park to see females den with their cubs. See www.frontiersnorth. com. Note, polar bears are dangerous and can attack without warning: always visit as part of an organised tour.

SNORKEL WITH WHALE SHARKS
Tanzania
Despite their vast size (they can grow to 40ft, or 12m), whale sharks display immense grace and gentle curiosity. Don't dive? Not a problem. You can snorkel with these beautiful cetaceans during the annual whale shark gathering (from October to February) near Mafia off the coast of Tanzania. Swimmers can see pods of male juveniles gather to feed, just a few metres from shore.
HOW Mafia Island can be reached by air from Dar es Salaam and Zanzibar. Guided trips by boat visit Chole Bay and the outer reefs; drift and wall dives are possible outside the bay area. See www.diveworldwide.com.

COMMUNE WITH CALIFORNIAN REDWOODS
USA
Scaling the heavens, the tallest redwood is a staggering 379ft (115m) tall. But it's the gnarly girth of these soul-stilling trees that also impresses; too wide to embrace but soft and springy to touch, their red fibrous trunks induce tree-hugging in the most jaded traveller.
HOW Northern California's Muir Woods, Humboldt State Park and Sequoia National Park are home to the tallest, oldest trees, but the USA has close to 100 parks preserving redwoods. See www.nps. gov/state/ca.

HAVE AN EPIC TRAVEL FAIL

Missed your flight? Forgot your visa? Meant to ask for ice, but instead you asked for sex? Don't worry. Travel fails are annoying and sometimes embarrassing, but they provide handy skills moving forwards.

Screw-ups happen despite your best-laid plans. The hip hotel you booked online turns out to be attached to a strip club. Your luggage ends up on a different continent. You're lost with no idea where your hotel is. Anxiety reigns in the moment.

And then you figure it out. The world doesn't end when your suitcase goes to Sydney, Nova Scotia and you're in Sydney, Australia. It's irksome and it takes extra effort to restock, but look at the side benefits such as realising how few items you really need on the road. Travel fails often have a silver lining.

Go ahead. Have the biggest, baddest travel fail imaginable. Head to the wrong airport. Tell the waitress you'd like a plate of socks with cheese. You can't go wrong because failure builds resilience and that makes you stronger and more adept at handling debacles in the future. JK Rowling has said that failure taught her things she could have learnt no other way. James Joyce described failures as 'portals of discovery'. Both are apt descriptions for what you gain from travels gone awry.

Not to mention travel fails make great stories to tell in the pub afterwards.

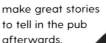

MISS YOUR FLIGHT

If you're going to flub it, do so in an airport with entertainment options. Amsterdam's Schiphol Airport has a gallery of art from the Rijksmuseum. Hong Kong's offers indoor golf and IMAX movies. Singapore's Changi Airport has a butterfly garden, 24-hour cinemas and city tours, all for free.
HOW Aerotel Airport Transit Hotel at Changi Airport's Terminal 1 has a rooftop pool with bar and Jacuzzi.

LOSE YOUR LUGGAGE

A visit to the Unclaimed Baggage Center in Scottsboro, Alabama shows you're not the only one. All US airlines send orphaned bags here, and the centre sells their contents at low prices, from racks of chino shorts to snowboard helmets and taxidermied frogs.
HOW Peruse the Unclaimed Baggage Center (www.unclaimedbaggage.com) Monday through Saturday.

GET LOST

In some destinations, going off course is a boon, leading to hidden courtyards and secret teashops. London, Venice and Istanbul are prime places to lose your way. Meanwhile, Mongolia recently adopted the digital address system called what3words that labels everything via 3m by 3m squares of land, so you won't get lost any more.
HOW Download the what3words app, type in the three-word 'address', and the location appears on your device.

MAKE A FOOL OF YOURSELF

Be proactive about it. Yes, you'll look silly samba dancing at the street festival or Chinese yo-yo'ing at circus school, but it can be liberating to suffer embarrassment and live to tell the tale.
HOW Give a speech at Speakers' Corner in Hyde Park, London. Anyone can rant in its northeastern corner on Sunday afternoons.

"I am a desk"

"I was feeling pretty cocky when our bus pulled up to the border post between Chile and Argentina in Tierra del Fuego. The lonely spot amid golden plains and a hulking ice-blue sky had a cowgirl vibe that spurred me on. It was also that I'd been in the region for three weeks practising my Spanish, which was getting good. Or so I thought.

I walked to the border agent's desk, passport in hand. He rattled off the usual questions. Where are you going? For how long? I threw down my best español. He barely looked at me. What is your occupation?

'Soy escritorio.'

With that his eyes flicked up and met mine. He smiled enormously. 'Welcome!' he said and stamped me into the country.

I glided back on to the bus. That's when Francisco, a fellow traveller who'd been next to me in the border queue, kindly explained, 'Escritora is writer. Escritorio is desk.'

My confidence deflated. I had just told the border agent I was a desk! And he wasn't the first I'd said it to. While it was embarrassing and humbling, it was also a lesson that people are mostly good-natured when you make a fool of yourself. Plus, I learnt two words that day that are seared into my brain forever."

Karla Zimmerman

TAKE IT FURTHER

Get to know your own prejudice:
Challenge your perceptions, p210
Find then stretch your limits:
Test your mettle, p230

JOIN A CONSERVATION PROJECT

Worried about climate change and the fate of our natural world? Be part of the solution, not the problem. Find or create a conservation project, and do your bit to protect and nourish our fragile planet.

Globetrotting travel can often feel like a hedonistic, and headlong, pursuit. Helping out on a local conservation project means you can give something back to the place you're travelling through, and it also gives you the benefit of slow travel: of seeing a small corner of the world in vivid detail. Whether you're getting your hands into some soil, counting animals in the wild or helping protect a coral reef, the rewards can be immense. As well as immersing you in a particular environment and eco-system, many conservation initiatives have community engagement and teamwork at their heart. Global friendships often result, and getting to know a place and its people close up can enhance your awareness of wider environmental issues and the problems facing our planet. Projects in remote environments with physically tough or repetitive tasks can really test your body and your emotions. However, despite the challenges, this type of volunteering is rarely a one-off in a volunteer's life. Most likely you will return home with improved skills and fitness and a hands-on approach to your own environment. You may find yourself inspired to grow bee-friendly plants on your balcony, or even to make a whole career in conservation. And all of us can benefit from living a simpler life for a few days or weeks. Don't be surprised if your whole attitude to the supposed essentials of modern life shifts, and your small step to help the planet turns into a stride.

BECOME A CITIZEN SCIENTIST

Biosphere Expeditions runs global projects from Malawi to Sweden, where volunteers assist scientists with data gathering in wild places. A reasonable level of fitness is required, but there is no selection process. You don't need prior scientific knowledge either, and with experts on hand to guide you, you'll soon feel like a budding David Attenborough.

HOW Do some physical prep for your trip: the fitter you are, the more help you'll be. www.biosphere-expeditions.org.

PLANT TREES

One Tree Planted aims to make it easy for you to contribute to the forests of North America, Latin America, Asia and Africa. Reforestation partners in these regions manage individual projects, such as the one-million-tree scheme, which was put into operation in California following the wildfires of 2017.

HOW Individuals can get guidance from the charity on funding and organising local tree-planting initiatives. See www. onetreeplanted.org.

TRAIN IN COASTAL CONSERVATION

Blue Ventures has developed a unique and successful model: it aims to transform coastal environments by incorporating social enterprise into conservation projects, encouraging fishing communities to improve their practices and thus increase their yields. The result is cleaner waters and recovered fish populations.

HOW Volunteers head to Madagascar, Belize or Timor-Leste to learn the skills that support the organisation's aims; www.blueventures. org.

Left: Plant a sapling with One Tree Planted

Red kites: back from the brink in the United Kingdom

With their sweeping, near 6.5ft (2m) wing span, distinctive forked tails and habit of catching a lazy lift on air currents, red kites are among the UK's most beautiful birds. Reduced to just a few pairs in Wales back in the mid-1980s due to egg-hunting, the birds were reintroduced to the west of England and central Scotland in the 1990s. Different populations, brought from Sweden, Spain and Germany, began to inter-breed and were encouraged to thrive with feeding centres. Now, some 20 years later, the UK has an abundant population of these russet red beauties. The scheme is counted as one of the most successful avian reintroductions in Britain.

TAKE IT FURTHER

Slow down and focus on a new skill:
Learn a craft, p94
Enjoy sustainable community living:
Go off grid, p192

On a planet plagued by light pollution, it takes effort to find areas far from the tangerine glow of urban centres. But the quest reaps a rich reward, as few experiences can connect us to our most primal place in the universe quite like seeing the majesty of the night sky.

There is magic in standing outside under a blanket of twinkling celestia. The stars impart a compelling sense of perspective about humanity's place in the universe: we are at once giant and tiny.

But according to research conducted by the Light Pollution Science and Technology Institute, 80% of the earth's land mass suffers from light pollution. Perspective-seekers must put in effort to reach places that are dark enough to be able to behold the cosmos in all its glimmering splendour.

Several national governments and organisations, such as the International Dark Sky Association, have begun awarding designations to places that are working to preserve the quality of their dark skies. But all the dark-sky-friendly LED lighting in the world can't fix the lingering horizon-glow from a nearby city. To really get dark, you need to seek, travel and even hike your way to the remotest corners of the planet.

And once at your destination, you may need to overcome extreme weather and wild terrain. As a reward, stargazing develops patience and concentration. A digital detox is also obligatory: for the human eye to take in small amounts of light, the pupil must dilate and adjust, so phones must stay in pockets. This is a time to contemplate, meditate and ruminate, and in doing so, connect to our deepest selves.

VISIT DARK PLACES

TAKE IT FURTHER
See mind-bending natural phenomenal:
Feel out of this world, p162
now your place in the universe:
Watch a total solar eclipse, p226

Left: Aoraki Mackenzie Dark Sky Reserve, New Zealand

CHACO CANYON NATIONAL PARK
New Mexico, USA
Between the 9th and 11th centuries, a remarkable culture in what is now remote northwest New Mexico built a series of stone forts along lines corresponding to the night sky. Accessible by dirt roads, Chaco Canyon is a designated dark-sky park with an observatory and stargazing events.
HOW Albuquerque airport is 150 miles (240km) away; from there you'll need your own car. Heritage Inspirations (www.heritageinspirations.com) runs glamping tours.

AORAKI MACKENZIE DARK SKY RESERVE
New Zealand
Centring on New Zealand's highest peak, Mt Cook, the world's largest dark-sky park covers a remote swath of the South Island. There are two observatories and several tours get you off the beaten track for either naked-eye or observatory-based stargazing.
HOW Mt John Observatory is three hours' drive from Christchurch; stay in Lake Tekapo village. Earth and Sky (www.earthandsky.co.nz) runs coach tours that include telescope-based observing.

LA PALMA
Canary Islands
The whole island of La Palma is a starlight reserve and is home to the Roque de los Muchachos Observatory, which has some of the largest telescopes in the northern hemisphere. At 7870ft (2400m), seated off the west coast of Africa, La Palma's skies are dark year-round. There are walks designed for night-time hiking, with stargazing viewpoints along the paths.
HOW The observatory (www.starsislandlapalma.es) is open in the day only; hiking trails are open year-round. Stay in nearby Garafía.

EAT
INSECTS

What do grasshoppers, crickets, beetles, butterflies, bees, ants and mealworms have in common? Yes, they're all insects and they're also really tasty! Even better, eating them is good for your own health and that of the planet.

*D*espite resistance in the West, where bugs are generally shunned as pests, eating insects, aka entomophagy, is common practice for more than two billion people in 100-plus countries. In fact, there are some 2100 identified species of edible insects and most pack significant nutritional value. Generally speaking, these high-energy nibbles provide equal or greater yields of protein, omega-3 fats and numerous minerals than equivalent weights of most major meats.

Just as critically, farming arthropods is quite sustainable, requiring less space and fewer resources than vertebrate animals – only 1 gallon of water per pound of crickets versus 2000 gallons of water per pound of beef – generating a fraction of the waste and greenhouse gases associated with livestock, and serving as a low-capital, low-tech earning opportunity, even in countries that have large poor populations.

So is it any surprise that given growing human populations, land scarcity and climate change – there may not be enough meat protein to feed the world in 2050 – edible bugs are generating a new kind of buzz? And not just scorpions on skewers. How about bug-based energy bars and crisps, cricket-flour pastries and insect-infused bitters? Delicious! No joke: bee brave.

OAXACA
Mexico

There are more than 600 varieties of edible insects in Mexico alone, but Oaxaca's *chapulines* (grasshoppers) are the most famous: boiled, sautéed, seasoned and sold everywhere. Los Angeles, California, known for its large Oaxacan population, also serves *chapulines* in many eateries, including Guelaguetza (www.ilove mole.com), perhaps the city's top Oaxacan restaurant.
HOW Fresh *chapulines* are available year-round, but best during Oaxaca's summer rainy months (June to September).

ROT FAI MARKET
Bangkok, Thailand

Although edible insects are found in abundance all over Southeast Asia, bug vendors at Bangkok's night markets promise particularly rich pickings. The usual insect suspects include deep-fried bamboo worms (called *rot duan* in Thai), crickets, locusts and scorpions, all made extra tasty with a little spray of soy sauce.
HOW Rot Fai Market in Ratchada, one of Bangkok's excellent night markets, is a good one for palate challenges. It's near the Cultural Centre MRT underground station.

GRUB KITCHEN
Pembrokeshire, Wales

For more formal dining, a growing roster of restaurants, including some of the world's best, uses bugs as both unseen and very visible ingredients. Grub Kitchen, which claims to be the UK's first insect restaurant, has a palate-expanding menu with bug-based sides and main dishes, as well as non-insect fare.
HOW Located in Pembrokeshire, Wales, Grub Kitchen (www.grubkitchen.co.uk) works in parallel with The Bug Farm, a visitor attraction, learning centre and working farm.

Left: Grasshoppers contain a similar amount of protein as chicken

© KIDSADA MANCHINDA | GETTY IMAGES

Be mindful, even about bugs

Joseph Yoon, a chef and executive director of Brooklyn Bugs (www.brooklynbugs.com), which works to raise awareness of and appreciation for edible insects, believes quality ingredients and responsibly sourced supplies are critical to the entomophagous equation. 'Let's say you have the most beautiful steak. If it's not good meat, it's not going to taste good. So when you're trying to work with edible insects, you want to find resourceful, but also responsible, vendors for everything. That's very important. It's something I really can't encourage more for everybody – to be more mindful, more respectful. Everything has an impact on the world around us.' Even bugs!

TAKE IT FURTHER
Challenge your carnivorous needs:
Go meat-free on the road, p236
Reduce your impact:
Live off the land, p272

GET LOST
IN A CROWD

Wild open spaces are all well and good, but we're pack animals at heart and there's a buzz to be had from being in the thick of the action, a sensory rush that comes when you feel like a member of the madding crowd, in the midst of a magical moment.

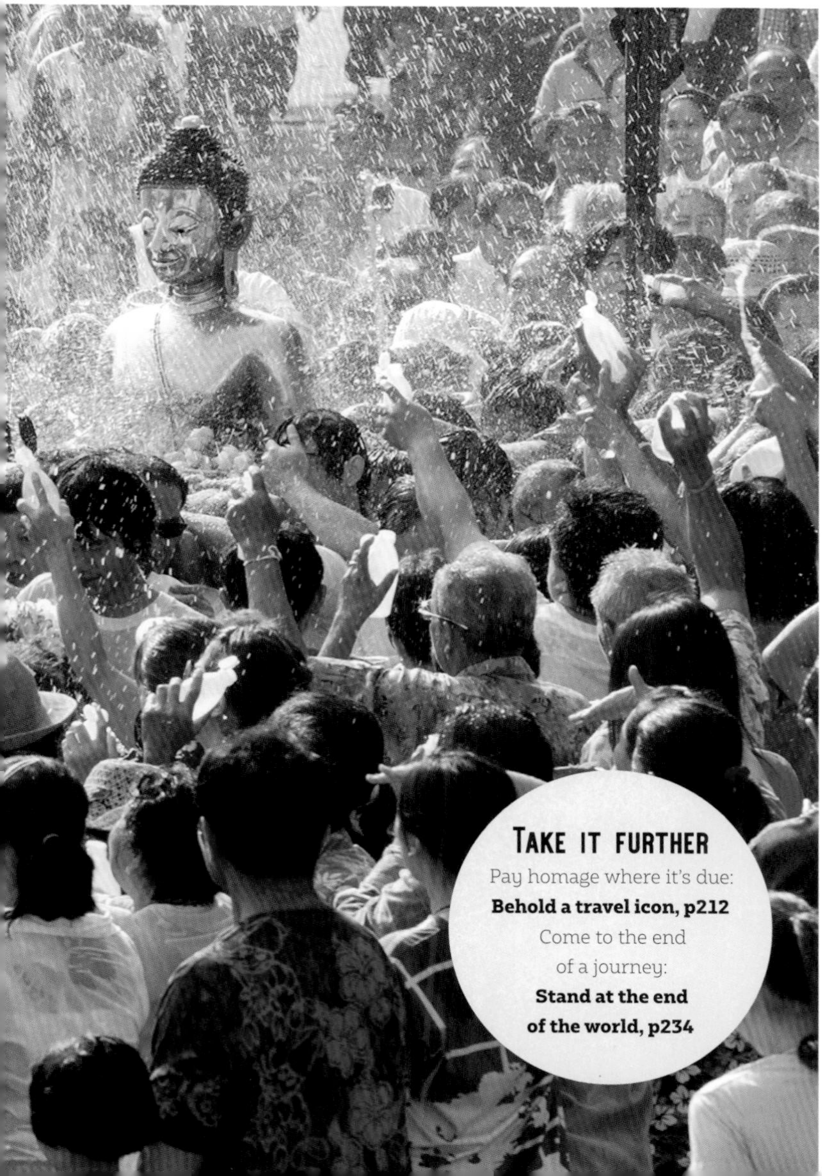

Take it further
Pay homage where it's due:
Behold a travel icon, p212
Come to the end
of a journey:
**Stand at the end
of the world, p234**

© SUKPAIBOONWAT | SHUTTERSTOCK

GARE DU NORD
Paris, France
International transport hubs are intoxicating places, as myriad human stories unfold and destination boards flicker with far-flung cities. Every year, 180 million travellers bustle through Gare du Nord, Europe's busiest railway station.
HOW See Gare du Nord at its hectic height on Saturday afternoon.

CARNAVAL
Rio de Janeiro, Brazil
Attracting two million people per day to the streets of Rio, the biggest and most-famous carnival in the world dates back to 1723. More than 200 samba schools (each representing a neighbourhood) shake their sequin-clad thing during the six-day shindig.
HOW Hotels, particularly in Rio's South River area, fill quickly. Book three to four months in advance.

MELBOURNE CRICKET GROUND
Australia
The MCG is one of the world's largest cricket stadiums and also hosts the Grand Final of the Australian Football League each September, when 100,024 fans pack the ground and the air crackles with anticipation and colourful language.
HOW Tickets for all MCG events are sold through Ticketek (www.ticketek.com.au).

SONGKRAN FESTIVAL
Thailand
During the Thai New Year, everyone takes to the streets for a massive water and paint fight, using water guns, filled balloons, buckets and anything else they can grab. In Bangkok, it's pandemonium.
HOW Songkran is celebrated countrywide on 13 April. In Bangkok, head to Khao San Rd for the best action.

Left: Songkran Festival being celebrated in Bangkok, Thailand

SEE A ROCKET LAUNCH

CROSS A REMOTE BORDER

RETRACE YOUR ROOTS

GO OFF GRID

SUMMIT A MOU

T EMPTY PLACES

FEEL OUT OF THIS WORLD

EXPERIENCE CULTURE SHOCK

SEE A ROCKET LAUNCH

TRAVEL BY BIKE

ENGAGE ALL YOUR SENSES

TAIN

4

SUMMIT A MOUNTAIN

Have a peak experience – literally – by climbing all the way to the top. You'll earn a sense of achievement, both physical and emotional, and get some killer views along the way.

There's a reason so many metaphors have to do with mountain tops: high point, pinnacle of life, peak condition. Getting high in the sky makes us feel powerful. It's not just about the climb, although most mountain summits involve vigorous physical exertion. It's also about the atavistic sense that being high up gets us close to something important – the stars, the gods, the mystery of life itself.

Mountains figure heavily in our collective legends and religious stories. Abraham followed God's word up Mt Moriah. Moses climbed Mt Sinai and came down with the Ten Commandments. The Greek gods lived on Mt Olympus. Mythical Mt Meru is the holy five-peaked mountain of Hinduism, Buddhism and Jainism. Climbing mountains is still a part of many modern religious pilgrimages, where sport meets the sacred.

Climbing a mountain may take more planning than your average beach trip – bagging an important peak often involves permits, guides and specialised equipment – but that's part of the fun. You don't have to summit Everest to make the journey worthwhile. There are plenty of impressive peaks accessible to the climber of average physical condition, and many that can be climbed in a day rather than a week.

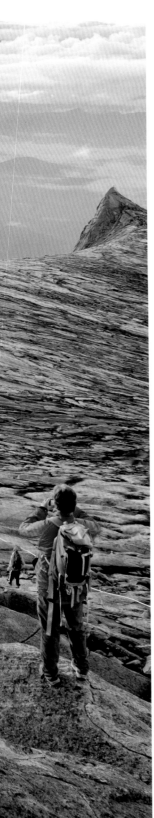

THE HALF-DAY: CROAGH PATRICK
Ireland
It's not the highest peak in Ireland, but County Mayo's Croagh Patrick, also called 'the Reek', holds an important place in Irish Catholic history. According to legend, St Patrick prayed on its summit for 40 days and 40 nights; pilgrims still climb the mountain barefoot. Non-believers gain both a quality quad-burn and jaw-dropping views across the islands of Clew Bay.
HOW Best done as a self-guided hike. There are numerous accommodations in Westport, 6 miles (10km) away.
Elevation: 2507ft (764m); time to top: two hours; best time to go: May–September.

THE FULL-DAY: VOLCÁN CONCEPCIÓN
Nicaragua
In a country of volcanoes, fire-bellied Concepción is one of the fiercest. It's half of the barbell-shaped island of Ometepe in the middle of Lake Nicaragua, accessible only by boat. Press through dry tropical forest, your climb punctuated by the screams of howler monkeys, all the way to the exposed, sulfur-scented summit, where you can barely see your own feet for the clouds of steam.
HOW You need a guide to summit; try Ometepe Unique Tours (www.ometepeuniquetours.com).
Elevation: 5282ft (1610m); time to top: five hours; best time to go: May–September.

THE MULTIDAY: MT KINABALU
Malaysian Borneo
Ascend through the lush Sabah rainforest to emerge on the sheer granite balds of this holy Malaysian mountain. You'll spend the night in a communal lodge before rising in the darkest pre-dawn in order to make the summit push. Pull yourself along slippery sheer rocks by holding on to chains as cold and altitude stiffen your hands. You'll be rewarded by sunrise above the clouds.
HOW Guided climbs must be booked well ahead via the park's official website: www.mountkinabalu.com.
Elevation: 13,435ft (4095m); time to top: 1.5 days; best time to go: February–April.

© YVES ANDRE | GETTY IMAGES

Left: The granite expanse of Mt Kinabalu

Everest Base Camp Trek

Let's be real: you're not climbing Mt Everest. But that doesn't mean you can't get a glimpse of the world's highest peak. Take the Everest Base Camp Trek, a popular two-week hike from the town of Lukla through the Sagarmatha National Park and the frozen landscape of the Khumbu. It's a route sometimes called 'the steps to heaven' for its otherworldly beauty. The trek takes you right into the high-altitude heart of the high Himalaya, more so than any other teahouse trek. You'll pass Sherpa villages and high-altitude monasteries with fluttering prayer flags, cross ravines on swinging bridges and perhaps spot native animals like musk deer, black bears and snow leopards. You'll need to be fit, but don't worry if you're no ultramarathoner. You can go it alone or with a guide company, of which there are many – such as Adventure Treks Nepal (www.adventurenepaltreks.com).

TAKE IT FURTHER
Get to know a place intimately well:
Cross a country on foot, p214
Harness the planet's power:
Conquer nature, p276

RETRACE YOUR ROOTS

Travelling back to your ancestral origins indulges both a need for a journey and for a rooted sense of self. And thanks to developments in genetic testing, it is easier than ever to pack a bag and make this often emotional odyssey.

Our families all come from somewhere, and for some people, 'where' is integral to a sense of who they are and what they can be. Ancestry exploration has swelled in recent years thanks to widespread DNA testing and more accessible genealogy records. Descendants of diasporas in particular may find this sort of research appealing: when your ancestors (or parents) were unwillingly exiled, there is a desire to see the home they never wanted to leave. The spicy food your friends wrinkle their noses at is suddenly the default cuisine; the music that your parents or grandparents listen to is no longer an oddity – it's on the radio station; the little tics and quirks you once thought, well, a little quirky, are behavioural norms; and the language you speak with your family is on the street, in the newspapers, and broadcast over the airwaves.

When undertaking this kind of journey, it is important to manage expectations. Travellers who are re-tracing family roots can come into the project with big ideas and big hopes – and sometimes this leads to big disappointments. After all, the place where your ancestors came from is not where you come from; it may feel welcoming and warm, but there is a chance that it will feel alien, even unapproachable. A good rule of thumb for heritage tourists to follow is to not project too much of their own desires on to a place. Whatever happens, though, it promises to be a life-changing experience.

AFRICAN HERITAGE IN GHANA

The Transatlantic Slave Trade was the largest deportation in history. From the 16th to 19th centuries, some 12 million Africans were sold into chattel slavery. Several tour outfits offer comprehensive, often harrowing, slavery heritage tours.

HOW Ashanti African Tours (www.ashanti africantours.com) and Jolinaiko Eco Tours (www.joli-ecotours.com) can set up heritage tours, with a focus on Ghana. Ashanti's tours can be tailored for length and cost, and include African naming ceremonies and visits to the forts where slaves departed the continent.

JEWISH LEGACY IN POLAND

Poland once contained the world's largest Jewish population. Jews were a vital cultural force in Poland and other Central and Eastern European nations until they were almost wiped out during the Holocaust. Today, tour operators can take visitors to Polish synagogues, Jewish cemeteries, the remains of ghettos and other sites; many tours focus on Warsaw, Kraków, and Lublin.

HOW Taube (www.taubejewishheritage tours.com) and Tauck (www.tauck.com) offer Jewish heritage tours in Central and Eastern Europe. Taube's one-week tour takes in major sites of Ashkenazi cultural heritage in Warsaw and Kraków.

Below (left and right): Portraits of Polish Jews at the Museum of the History of Polish Jews, Warsaw; Elmina Castle, Ghana, one of the most significant stops on the route of the Atlantic slave trade

RETRACING IRISH ANCESTRY

The Irish diaspora occured in several waves and for several reasons, not least of which was the Great Famine of the 1840s. Today, the descendants – which some count at 80 to 100 million individuals – can be found in countries from the UK to India and the USA to New Zealand. Irish heritage tours are not limited to large cities and can retrace ancestry to the village level.

HOW My Ireland Family Heritage (www.myirelandheritage.com) claims to find ancestors' specific house locations and runs up-to-date family trees.

© ANDIA | ALAMY STOCK PHOTO. © INIGO BUJEDO AGUIRRE | VIEW | SHUTTERSTOCK

Shwedagon sojourn

"My Burmese mother left Myanmar willingly, but her homeland was always a storybook to me, a land of elephants and monks and mist and flowers … and family – my Burmese grandparents lived in Yangon while I was in America. I first went to Myanmar, then known as Burma, when I was 17. It was one of my first major trips abroad, and the spoiling I received from my extended Burmese family, coupled with my own wide teenage eyes, did nothing to dissuade my romanticism.

A I grew older, I started travelling to Burma – which was becoming Myanmar – at least once every two years. I was lucky, not just to be able to travel, but to couple my heritage tourism with visits to living family members. As the years went by, some family members died. Others said things I deeply disagreed with. Myanmar democratised – hope! – but this was followed by the displacement and murder of the Rohingya people – tragedy. Despite all of the above, I still find myself utterly spellbound by Yangon's Shwedagon Pagoda at sunset: a mass of rustling robes and whispered prayers and burning candles. For me today, Myanmar represents neither my schoolboy idealisations nor their reverse, but a real place, complicated and messy. And it has only been through many trips that I have realised just how little I know about it."

Adam Karlin, travel writer

TAKE IT FURTHER

Come home a different person:
Change your identity, p220
Exercise your self-awareness:
Find yourself, p274

*N*othing stokes the travel buzz like crossing a far-flung border. You know the type: a true outpost where the action happens in a shed by the side of the road. You enter the little building, and the lone guard with a heavy gun holstered at his hip takes your documents. He eyeballs your passport, then you, then your passport again. He jabs a finger at one of the pages and says something you don't understand.

Your heart beats hard. You're pretty sure your paperwork is in order. You're also pretty sure there's no one out here to help if this meeting goes awry.

The guard frowns and takes a banknote from his pocket. It seems you forgot to pay a tax? An entry fee? You slide a similar bill across the table. He thwacks a passport stamp into your book. And you're on your way!

The world is full of weird, wonderful and sometimes edgy crossings. Consider the curious border between Attari, India, and Wagah, Pakistan, where moustachioed guards meet each evening to close the post in a bizarre ceremony of competitive marching, flag-folding and high-stepping that's equal parts military showmanship and theatre. Or ponder the chill border at the Northwest Angle between Minnesota, USA, and Manitoba, Canada, where the DIY crossing is a minuscule shack with a videophone inside, on which you ring a distant customs agent and ask for permission to enter. It's a far cry from the Kyrgyzstan–China border, with formalities aplenty but also mountaintop views and Silk Road caravanserai to add to the adventure.

CROSS A REM

TAKE IT FURTHER

Find magic in the mundane:
Appreciate the ordinary, p200
Travel to the end of the earth:
Step foot on the frozen continent, p250

KHUNJERAB PASS

China–Pakistan
Superlative seekers wanting to cross the world's highest border will find it at 15,500ft (4725m) along the fabled Karakoram Hwy. Spiky mountains, herdsmen with yaks and ever-falling snow provide the eye candy between border posts at Sost, Pakistan, and Tashkurgan, China.
HOW Go May to December; the border closes the rest of the year due to dangerous weather conditions. Organise visas for both countries in your home country.

SANI PASS

South Africa–Lesotho
Prepare for high drama at this frontier. The South African border is at the bottom of the pass, while the Lesotho border sits at the top, dotting a mountain. A crazy-steep, hairpin-twisted road rises 4364ft (1330m) in 5 miles (8km) between the two, offering spectacular views of rugged cliffs, waterfalls and blanket-cloaked shepherds en route.
HOW Best November to March when the weather is dry. A 4WD vehicle is required; hire one in Durban, or go with a tour such as Sani Pass Tours (www.sanipasstours.com).

PONTA PORÃ

Brazil–Paraguay
It's a strange feeling to cross the street and be in a different country, but that's how it works in Ponta Porã and Pedro Juan Caballero, two towns joined at the hip where the dusty main road through the centre is the international border. The beer changes brands, hot maté switches to iced maté, and real bills swap to guaraní bills from one side of the street to the other.
HOW Get your passport stamped by the local police on both sides.

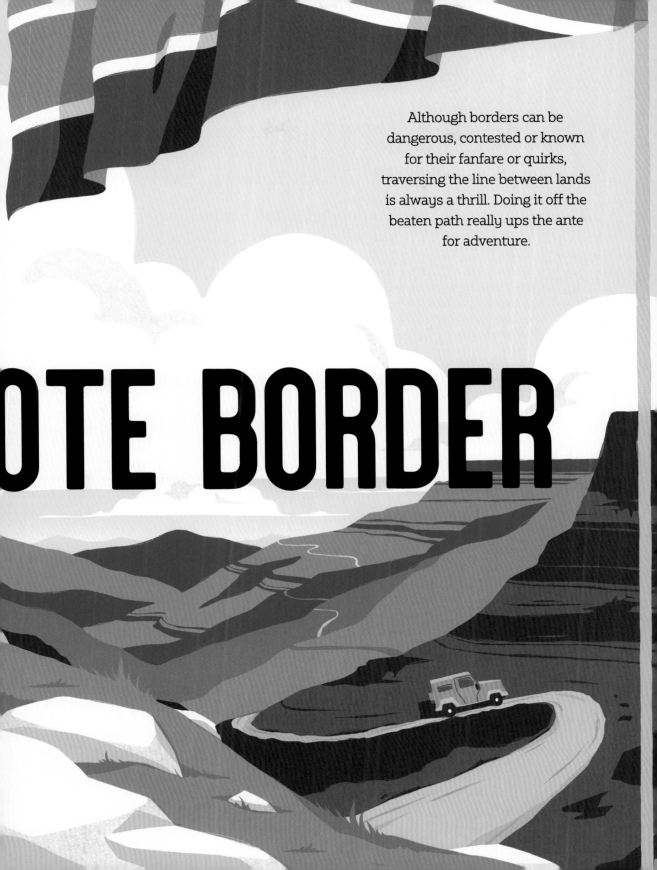

Although borders can be dangerous, contested or known for their fanfare or quirks, traversing the line between lands is always a thrill. Doing it off the beaten path really ups the ante for adventure.

OTE BORDER

WRITE A TRAVEL BLOG

Hone your writing skills and record your travel adventures. Writing a travel blog is a fun, easy way to process and remember your experiences, as well as to share your stories with friends near and far.

The benefits of keeping a journal, whether home or abroad, are well documented: the daily practice of writing thoughts and experiences can help reduce stress and inspire creativity. Journaling while travelling is particularly useful, since our travel days are dense with new people, places, sights and sensations. Sometimes we pack so much in, it can be difficult to remember everything we do. Taking photographs is good, but the photos are even better when they accompany a first-person narrative, recounting our impressions of our experiences as they occurred.

A blog shares many of the benefits of a journal but takes it to the next level. This is a public forum, so writers should be more selective about what they post and how it's written. For this reason, a blog can be a source of stress instead of a stress reducer. That said, the process of crafting and refining a text – creating something 'good enough' for others to read – is a valuable practice for processing experiences and improving writing skills. It takes time, commitment and courage, but the result is priceless: a written record that allows us to share our stories with our friends, family and future self.

BACK TO SCHOOL

The basics of travel blogging seem easy enough (travel, write, repeat). But if you intend to make money from this gig, a lot of factors can make or break your blog. Learn from the experts on how to establish a successful blog.

HOW Try Superstar Blogging (https://superstarblogging.nomadicmatt.com) by Matthew Kepnes, aka Nomadic Matt.

MAKING IT A HABIT

So you went on vacation and started a travel blog. Congratulations! Now how in the world are you going to sustain this thing? Take a stay-cation and become a tourist in your home town, uncovering little-known destinations or sharing other insider tips.

HOW Set a goal to post at least once or twice a week and keep a running list of topics to fall back on when needed.

Below (left and right): Blogging regularly can help reduce stress

REFINE AND RETREAT

When you want to refine your craft, it's time to retreat. Free from distractions, surrounded by spectacular beauty and inspiring people, ideas alight and words flow freely. At least that's the plan.

HOW Your options are plentiful, but some recommended writing retreats include the Iceland Writers Retreat (www.icelandwritersretreat.com) and Wide Open Writing (www.wideopenwriting.com) held at locations around the world.

THE BUSINESS OF BLOGGING

Are you wondering how to use SEO to reach more readers? Are you thinking of monetising your blog or even becoming a full-time blogger? A few excellent industry conferences offer opportunities for bloggers to network, listen and learn.

HOW The biggest conference is TBEX (Travel Blog Exchange; www.tbexcon.com), which takes place annually in Europe and the USA.

The same trip, but better

"I have always been a journal writer, especially while travelling. But going public forced me to start refining my thoughts into cohesive stories, instead of just spewing them on to a page. When I started my first travel blog, I was terrified. I was already a published author, but this was so much more personal than anything I had ever written for publication. I didn't tell anybody about the blog for several months. Finally, I told my parents, and they forwarded the link to everyone they knew. (The lesson here is to not tell your parents unless you're ready for a much broader audience!)

Over the years, I have started and stopped and started again several travel blogs. For me, the value lies in recording and remembering my travel experiences in a way that I can also share them with others. But there's another more subtle benefit: the very act of writing and refining often enhances my understanding and appreciation of the experience I am writing about. So the blog allows me to get more out of the very same experiences."

Mara Vorhees, travel writer and blogger at www.havetwinswilltravel.com

TAKE IT FURTHER
Journey back in time:
Take a vintage transport method, p186
Make travel your lifestyle:
Don't stop travelling, p278

BECOME AN OC

How to be an ocean-friendly traveller

1. Dine consciously
Download the Seafood Watch app to make sure what you purchase was harvested sustainably. You'll need to know the type of seafood and location it was caught for the tool to work, so if you don't have that information, consider ordering something else.

2. Say no to plastic
Governments around the world are finally waking up to our plastic-choked reality. But as long as single-use plastic remains available we need to do our part and opt out. Carry your own reusable bags, water bottle and cutlery wherever you roam, and remember to tell your waiter to hold the straw.

3. Use a reef-safe sunscreen
Chemicals commonly found in sunscreen, such as oxybenzone and octinoxate, are now known to have a damaging effect on marine ecosystems, including coral reefs. Helpfully, with growing awareness, it's easier than ever to find a brand that actively avoids these ingredients.

4. Take three for the sea
Get in the habit of 'taking three for the sea'. Each time you hit the beach (or any waterway or natural area) don't leave until you've picked up at least three pieces of plastic trash.

5. Choose eco-certified travel operators
Make an effort to choose the most responsible marine tourism operators while travelling. The first step is to ensure that operators are licensed and their guides are certified. Ideally, operators will also hold a form of national or international eco-certification. If the information isn't available, don't be afraid to ask.

6. Get active
A growing number of conservation foundations (see right) run citizen scientist programmes that allow everyday travellers to play an active role in contributing to the ocean's long-term protection.

AN DEFENDER

The ocean is life. We rely on it for food, transportation and recreation. Even if you aren't a water person or seafood fan, you still depend upon it. The oceans, and their phytoplankton, provide half of the world's oxygen supply. It's responsible for our every second breath.

But you wouldn't know it based on how we treat the waters. You've seen the footage. The ocean has become our dumping ground for single-use plastic, among many other things. Warming temperatures are bleaching the last remaining coral reefs. Overfishing has decimated fish populations, and trawlers and long-line operations kill sharks, whales, dolphins, rays and turtles as bycatch. All hope is not lost, however. The ocean can regenerate. Whales were once hunted to the brink of extinction, but the majority of species have rebounded. Most fish produce thousands of eggs at a time. Coral will have a more difficult time, but successful transplantation and seeding projects are now happening on reefs in the Caribbean, the Maldives, Indonesia and Australia, among others. Time is running out. It is up to every single one of us who love the ocean to do everything we can to educate, change our behaviour and agitate humankind to protect what's left and spur recovery. Our lives could literally depend on it.

TAKE IT FURTHER

Commit to the lifestyle:
Travel without plastic, p204
Become educated about climate change:
Learn about fragile places, p244

Left: Volunteers cleaning the beach in Puerto Morelos, Mexico

5GYRES

5Gyres is one of the most internationally respected marine plastics research organisations, and you can help it spread citizen science and push healthy ocean policies by becoming an ambassador. You'll receive a tool-kit that you can use to hit schools and community events to spread the word. **HOW** Every summer, 5Gyres (www.5gyres.org) leads a crew of citizen scientists on an expedition to research a specific gyre (north Atlantic, South Pacific or even the Arctic). Participants help with microplastic sampling and beach clean-ups.

SURFRIDER

Another internationally renowned non-profit, the Surfrider Foundation is all about keeping our oceans clean and advocating for environmental policies. Wherever there are waves there is likely a Surfrider chapter and a beach clean-up on the calendar. **HOW** Surfrider (www. surfrider.org) chapters in many different cities host regular beach clean-ups. Check out the website to find one near you.

WATERKEEPER ALLIANCE

Combining detective work with advocacy, Waterkeeper chapters are staffed by locals who track pollution to its source, then document it to influence policy. Waterkeepers take care of rivers, lakes and bays, and they can always use a donation or a hand. **HOW** Los Angeles Waterkeeper (www.lawater keeper.org) has a marine protected area (MPA) watch programme that gives volunteers a chance to help survey and monitor human activities, and get up close to marine animals, in the MPAs around Santa Monica Bay.

FEEL OUT OF THIS WORLD

Some natural wonders feel more supernatural. Experience the mind-bending awe of some of the earth's weirdest and wildest phenomena, from bioluminescent beaches to the aurora borealis, and come away with a renewed sense of reverence for our crazy gorgeous planet.

*T*hings we didn't understand, we used to attribute to gods or magic. The Northern Lights were caused by the tail of an enchanted fire fox, or by the torches of the dead, or by legendary warriors battling in the sky. Yellowstone's geysers were the souls of animals killed by a mythical hero, while its hissing steam vents were the hot breath of a dinosaur slain when natives threw boiling water down its gullet. The lights of fireflies were the souls of soldiers lost in war.

The realities – collisions between particles in the atmosphere, hydrogeological forces, chemical reactions – are not as poetic, but no less beautiful in their own way. How marvellous is nature to evolve a creature like a firefly? How extraordinary that disturbances in the earth's magnetosphere should produce dancing curtains of green and purple?

Science aside, seeing wild and psychedelic phenomena transport you to a place of ancient awe. They make you feel small, yet connected. They bond you with your fellows through rituals of oohing and aaahing. And even though you know the truth, they make you wonder if you might just have glimpsed a fire fox, shaking his sparking tail across the night sky.

MOONBOW OF VICTORIA FALLS
Zambia

The rare moonbow, or lunar rainbow, is a phenomenon caused by moonlight shining through a fine spray of water, typically from a waterfall. One of the few places on earth to spot one is at Victoria Falls, on the Zambezi River. Catch it on a night with a full or near-full moon. The bow starts above the crest of the falls then lowers and brightens with the moon's arc.
HOW Take a two-hour Lunar Rainbow Tour with Zimbabwe's Wild Horizons (www. wildhorizons.co.za).

BIOLUMINESCENT BEACHES
Vieques, Puerto Rico

Trail your hand in the dark night ocean and see it awaken a ripple of tiny blue stars. This is bioluminescence, a phenomenon caused by single-celled organisms called dinoflagellates, which glow when disturbed due to an enzymatic reaction. One of the highest concentrations is off Mosquito Bay on the island of Vieques.
HOW Abe's (www.abessnorkeling.com) offers guided kayaking trips.

GLOW PLANTS
Hachijō-jima, Japan

During the May-to-September rainy season on Hachijō-jima, a tiny volcanic island off Japan's southern coast, more than half a dozen species of bioluminescent mushrooms glow green on the forest floor. Seeing them is a psychedelic experience.
HOW Embark on a night hike in the woods or visit the island's botanical gardens after dark. Regular flights from Tokyo to Hachijō-jima take about 45 minutes.

© KOLBEIN SVENSsON | 500PX | GETTY IMAGES

Left: Spot the Northern Lights (aurora borealis) in destinations near the Arctic Circle

Where to see the Northern Lights

Now understood to be caused by disturbances in the earth's magnetosphere, the dancing green, blue and purple lights of the planet's far north have long been a source of wonderment for humans. You can catch the magic from various spots around and near the Arctic Circle, from Alaska and Canada to Greenland and Scandinavia. Top spots include Canada's remote Yukon Territory and Norway's Finnmarksvidda plateau, with no mountains to obscure the eerie dance. Pick a cloud-free night, don your warmest parka and prepare for the head-to-toe tingles of seeing the sky swirl in shades of lime green and glowing indigo. December through March see the fewest clouds.

TAKE IT FURTHER
Feel small:
Watch a total solar eclipse, p226
Seek the beautiful and rare:
Witness a miracle of nature, p282

EMBRACE YOUR SEXUALITY

Coming out is seldom an easy or straightforward process, and for many LGBT+ people, travel is an ideal way to embrace their true selves. Attending a Pride parade or event is not only a chance to party but also to connect and identify with fellow travellers on the pan-sexual highway.

For those not sure of, or struggling with, their sexuality, heading overseas – or even within their home country – can be a liberating experience. Even if you grew up in the most accepting of environments, few people want to be the 'only gay in the village' so the appeal of places with active LGBT+ scenes is seductive. Locations such as London, New York and Sydney act as beacons for queer folk, so much so that in specific areas it can be a matter of being the only straight in the gay village.

This is not just about finding bars, clubs and hotels that are welcoming of all stripes in the sexual rainbow. Even in the most liberal of countries homophobia remains a hard fact of life. LGBT-support organisations around the world are therefore important refuges and crucial bastions in educating the general public and campaigning for human rights. The LGBT+ Pride events and parades that such organisations host are chances to honour the struggles of the past and celebrate the achievements of the present.

AMSTERDAM GAY PRIDE

The Netherlands

In 2001 the Netherlands became the first country to legalise same-sex marriage, and Amsterdam's gay scene is among the world's largest. On the first Saturday in August, the city's Gay Pride Festival culminates with a waterborne parade, in which spectacularly decorated boats packed with revellers float along the canals.
HOW For info on joining or just viewing the parade, see www.amsterdamgaypride.nl.

CHRISTOPHER STREET DAY

Berlin, Germany

Berlin has a well-deserved reputation as a counterculture and LGBT+ hotspot. The Berliner Christopher Street Day Parade (or CSD Berlin), the climax of the city's week-long Pride Festival, has taken place on a Saturday in late July every year since 1979.
HOW The parade route changes each year. See www.csd-berlin.de for details.

GAY PRIDE TAIPEI

Taiwan

Social tolerance is the norm in Taiwan and Taipei's four-day Pride festival, on the last Saturday in October, is one of Asia's largest LGBT+ events. The Pride parade starts and ends at the Presidential Office Building and politicians often show up to support (and court the votes of) the LGBT+ community.
HOW See www.twpride.org for event info and www.gaytaipei4u.com for where to stay and what to do in the city.

CAPE TOWN PRIDE FESTIVAL

South Africa

Late February is when Africa's pinkest city hangs out its rainbow flags and dusts off its feather boas for its annual Pride festival. Also bookmark December for the LGBT+ fancy-dress event MCQP.
HOW See www.capetownpride. org and www.mcqp.co.za for dates and event information.

Below (left and right):
Participant at the NYC Pride event; boats full of Pride in Amsterdam

© ANDREI ORLOV | SHUTTERSTOCK. © IVO ANTONIE DE ROOIJ | SHUTTERSTOCK

Coming out in Sydney

"My journey to embracing my sexuality as a gay man took me around the world, from my birthplace of Blackpool, UK – as LGBT-friendly a seaside resort as you could hope for – to Sydney, Australia. It was there in 1995 that I attended my first LGBT+ Pride parade. As I stood cheering on the floats, the dancers, the drag queens and the Dykes on Bikes, I made a vow that one day it would be me also doing the marching.

Seven years later that goal was achieved when I joined a group of friends who had secured a coveted slot in the parade. Weeks of rehearsals and careful preparation – including learning choreography, decorating our float, designing costumes (skimpy T-shirts and shorts) and securing a photographer and videographer to record it all – culminated in the big night itself. When I look back on it now I can still vividly recall the love of the cheering crowds and the enormous sense of pride I felt.

Travelling the world for my work, I go to places such as Russia and Iran where the LGBT+ community is stigmatised, discriminated against and forced into the shadows. This reinforces in me the continued need to work towards a future where we can all feel safe to be out and proud about who we really are."

Simon Richmond

TAKE IT FURTHER

Update your self-image:
Change your identity, p220
Dare to be authentically you:
Be yourself, p290

Turn holiday weight gain on its head, and use your next time abroad to get fitter, stronger and happier. This won't only pay dividends in the long run, but also when you're travelling – doses of exercise (and all those endorphins) will put a little more sunshine into your days.

ixing fitness with travel, or focusing your travel around it entirely, can be truly rewarding. For some this could mean hitting it hard at a fitness-focused resort to push your body to new highs, for others it may be completing a destination marathon, dedicating a week to learning a new sport or simply fitting in a few jogs before breakfast. The latter is something that almost everyone can do, and it actually makes for a great way to explore your destination. Running in the quiet, early morning through the streets of cities, whether New York, Rome or Tokyo, is a rather intimate experience and allows you to see another side to these usually buzzing metropolises. And a breakfast earned tastes so much the sweeter.

Having a focused holiday goal such as learning to surf or to rock-climb can be brilliant fun, and it also amps up your fitness and burns an ocean full (or mountain's worth) of calories. You may even find a new passion, and with these new skills in the bag you'll also be able to build them into future holidays or back at home.

Incorporating regular exercise won't just have the well-documented physical health benefits, but also mental ones, too. It's been shown to boost your overall mood, relieve stress and improve both memory and the quality of your sleep. And although less enduring, it's impossible to quantify the elation of pushing your body to a new personal best.

COME HOME FITTER

TAKE IT FURTHER

Experiment with stamina:
Test your mettle, p230
Make an effort for the greater good:
Pull together for the team, p252

SPORT AND FITNESS RESORTS

With Olympic swimming pools, running tracks, weight rooms, aerobic classes, HIIT (high-intensity interval training) sessions and CrossFit, these resorts are designed to elevate your performance. Certified instructors are on hand to provide support, and races are held (triathlons, duathlons, half-marathons etc) to test your mettle.
HOW Club La Santa (www.clublasanta.co.uk) on Lanzarote in Spain's Canary Islands is renowned globally for its facilities and instructors. It also offers road cycling, stand-up paddle boarding and windsurfing.

LEARN TO SURF

Your shoulders will scream, your abs will strengthen, and your smile will grow. First attempts tend to end before they start, but when you begin to get the hang of it the feeling is addictive. The cycle of exertion and elation will keep you going day after day and send you home fitter than you'd ever imagined.
HOW Australian Surf Tours (www.australiansurftours. com.au) uses video and skateboarding to push your fitness and skills.

YOGA AND ROCK-CLIMBING RETREATS

These two activities may seem at opposite ends of the sporting spectrum, yet they share many fundamentals including flexibility, strength, balance and mental focus. Combining learning yoga and rock climbing, or improving both if you're no novice, is incredible for both body and spirit. There are an increasing number of options globally.
HOW Climb Mediterranean (www.climbmediterranean. com) runs various retreats in Greece and Cyprus; Asturias Yoga (www.asturiasyoga. com) operates in Spain's El Parque Nacional de Los Picos de Europa.

Left: Rock climbing in Kalymnos, Greece

UPGRADE

Fly first-class, splash out on a suite, book the fanciest massage package at the spa. Sure, it's not an everyday thing, but that's what makes treating yourself unforgettable.

If you're always splurging on fancy things, it's possible you're either very rich or very broke. But even the fiscally responsible non-one-percenters among us can – and should – treat ourselves on occasion. The key is to understand what treating yourself actually is, and what's merely spending money.

It's the difference between racking up credit-card debt on souvenirs we can't afford and don't really care about and carefully saving up credit-card points to enjoy an anniversary night in a special hotel. Treating yourself can make ordinary moments extraordinary and turn already amazing adventures into never-forget-even-on-your-deathbed experiences. It can pull us out of dark psychological holes or simply shake up our everyday existence enough for us to appreciate the inherent wonder of life.

Treating yourself doesn't necessarily entail forking out a fortune – it could simply mean spending a day lounging in your hotel bed rather than seeing the sights, just because you feel like it.

Travel Goals

SAVOUR THE ULTIMATE OMAKASE
New York City, USA
Masa Takayama makes fish into artform every night on the hinoki-wood sushi bar of this NYC icon. The menu is *omakase*, Japanese for 'I leave it to you', meaning the chef serves whatever they want based on freshness, availability and creative whim. It's the most expensive restaurant in New York.
HOW Reservations are crucial; see www.masanyc.com for information. 4th fl, 10 Columbus Circle.

RENT A VINTAGE CONVERTIBLE
California
A trip down California's scenic seaside Route 1 is delightful in your dented 1998 Honda. But how much more delightful would it be in a 1965 Mustang convertible, wind in your hair? Some trips simply call for travelling with the top down.
HOW LA-based Legends Car Rentals (www.legendscarrentals.com) has convertibles from classic Porsche Speedsters to 1970s Cadillac DeVilles.

BOOK THE FANCY SEATS
Singapore
Cinemas in cities from New York to Hong Kong have been jumping on the luxury bandwagon, offering VIP cinemas with plush reclining seats, blankets and in-theatre food service. Head to the Gold Class theatres of Singapore's Golden Village cinema chain for leather recliners, chef-curated dishes like smoked duck salad, and a dedicated theatre concierge.
HOW See www.gv.com.sg/GoldClassHome for movie schedules.

STAY IN A ROOM WITH A VIEW
Canberra, Australia
No, we're not talking ocean view (though that's nice of course). For a rarer treat, splash out on something wild. At the Jamala Wildlife Lodge, at Australia's National Zoo in Canberra, guests can relax in their suite as lions and giraffes wander just beyond the window.
HOW Book a one- to three-night stay well in advance; see www.jamalawildlifelodge.com.au. National Zoo & Aquarium, 999 Lady Denman Dr.

Left: *Flying business class is the ultimate travel treat*

TAKE IT FURTHER
Know small gratitudes:
Appreciate the ordinary, p200
Take a lesson in self-care:
Meditate with masters, p216

How to bag a first-class upgrade

It used to be that you could just rock up at the airline counter in a fancy suit and slyly ask if there were any openings in business class. No longer. Airlines now hoard their best seats to reward customer loyalty, and unfortunately there are bound to be people who fly way more than you do. But all hope is not lost! Follow these tips for your best chance at an upgrade. Oh and that fancy suit can't hurt, either!

• Join a frequent-flyer programme. You may never make it to Ultimate-VIP-Million-Carat-Double-Diamond Class, but at least the airline can recognise that you travel with them regularly.
• Travel solo. It's rare enough to have an empty seat in First. Two or three? Probably not going to happen.
• It goes without saying, but: be nice. That means check-in counter staff, gate staff, flight attendants. You never know who might be in a position to bend the rules a little bit. Plus, karma.
• If you have special circumstances, let the staff know, nicely (see above). Are you pregnant? Is it your honeymoon? Are you 6'10" (2m)? It may not make a difference, but if staff have extra seats they might think of you first.

Don't let the bike gear and gears intimidate you. Embrace the generous benefits of cycling everywhere, no matter how near or far you go.

On November 2018, France's Robert Marchand celebrated his 107th birthday with a 9-mile (15km) bike ride. Two years earlier, at the sprightly age of 105, after setting a cycling record for his age class, he attributed his longevity in part to regular two-wheeling. Clearly, Marchand is exceptional. But for everyday bike riders, any cycling – travelling by bike over long distances or short daily commutes – is just as extraordinary.

Which is why it never hurts to set saddle-time goals, whether distance covered or time committed, exercise regime or holiday resolution, solo outing or group pursuit. A daily commute troop is a great way to break out of a rut and get (or stay) fit. More ambitious weekend rides with friends or clubs lead to eye-opening discoveries of your extended environs and physical resilience. From that, it's a few turns of the crank further to extended adventures across borders, perhaps even up (and thrillingly down) the categorised climbs of world-famous bike races.

Whatever the case, getting about by bike is both soothing and invigorating – a liberating, eco-friendly way to travel and live. The health benefits are considerable, too. A recent British study of the relationship between age and physiological function found older cyclists healthier than inactive adults. No surprise there. But the bikers were also found to be biologically younger. Wow.

Whether 17 or 107, it's good to stay bike spry. That might make 107 the new 70.

TRAVEL BY BIKE

NEW YORK CITY GREENWAYS, USA

For commuters, a growing gaggle of major cities has solid (and improving) bicycle infrastructure: marked paths, park trails and greenways that facilitate pedalling amid the urban sprawl. NYC's greenways, such as the 31-mile (50km) Manhattan Waterfront Greenway or the 40-mile (64km) Brooklyn–Queens Greenway, are some of the best.
HOW Resources, including maps, for NYC's bike network across all five boroughs and beyond are at www.nyc.gov/bikes.

LOIRE BY BIKE, EUROPE

European riverside bicycle trails are for all ages and abilities. Mostly flat, well paved and traffic free, they even touch the hearts of historical towns and cities. In France, the Loire path traces over 375 miles (600km) of castles, gardens and vineyards; its extension, as EuroVelo 6, parallels the Danube to the Black Sea.
HOW April and May are ideal, avoiding summer heat, holiday traffic and high-season prices. See www.loirebybike.co.uk.

TOUR D'AFRIQUE

A trip for the stout of heart and leg, this annual, continent-spanning bike race and expedition is epic in scope, touching about a dozen African countries on its 7500-mile (12,000km) journey from Cairo to Cape Town. Tour operator TDA Global Cycling offers Tour d'Afrique trips as well as similar epics across five other continents.
HOW The fully supported Tour d'Afrique ride departs Cairo every January and takes about 120 days to reach Cape Town. See www. tdaglobalcycling.com/tour-dafrique.

Below: Riding in the Rocky Mountains, Canada

TAKE IT FURTHER

Experience self-suffiency in wild places:
Go off grid, p192
Choose the slowest road:
Traverse a country on foot, p214

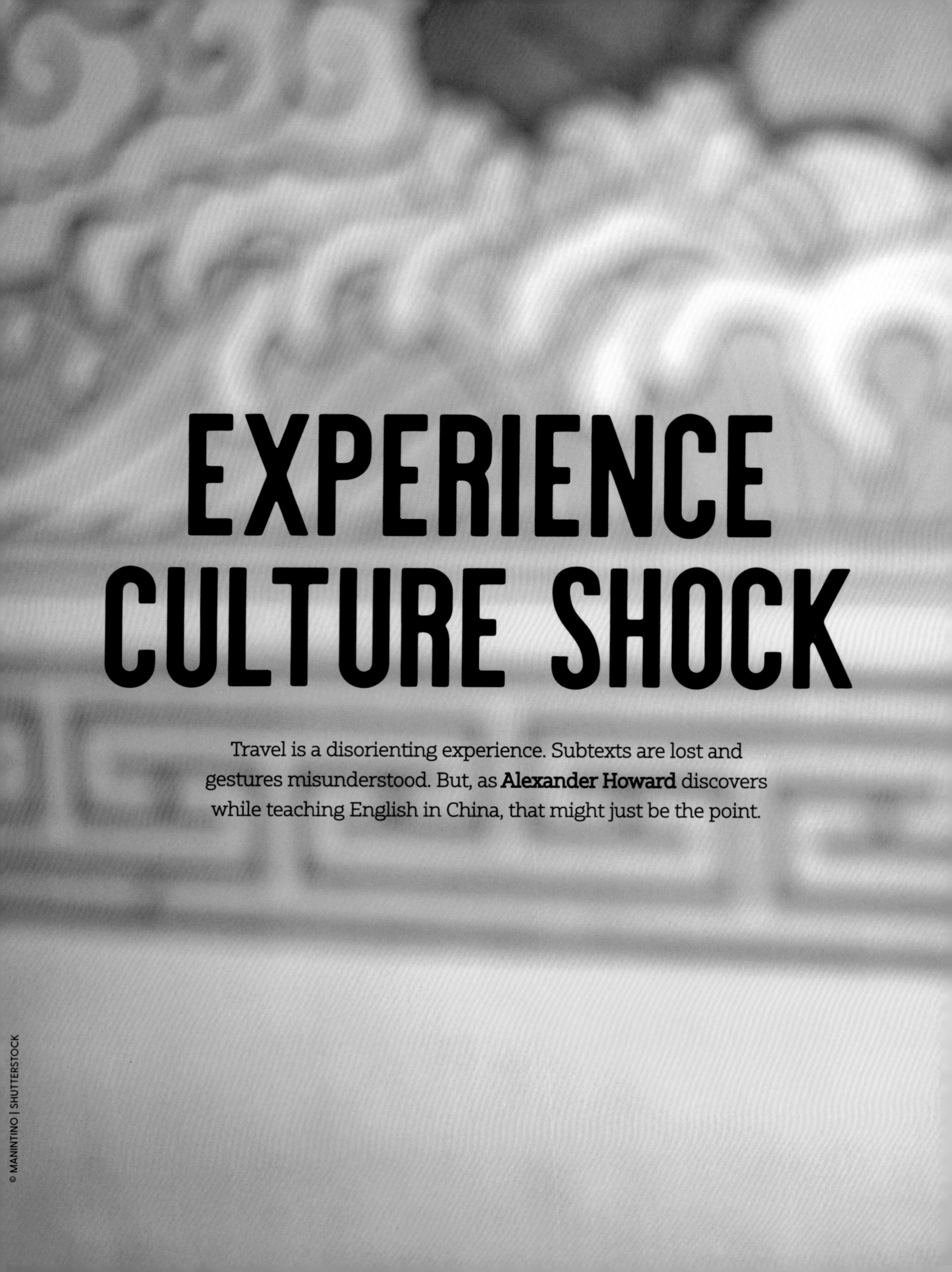

EXPERIENCE CULTURE SHOCK

Travel is a disorienting experience. Subtexts are lost and gestures misunderstood. But, as **Alexander Howard** discovers while teaching English in China, that might just be the point.

'You're getting a little fat,' my student Tina said to me. Other students filed out of the classroom, chatting gleefully with their friends on the way to their next class. I stopped erasing the chalkboard. 'Pardon?' I said, wiping chalk from my hands.

'Maybe you like Chinese food too much,' she said. And then she left, her message delivered.

I was in my third month of teaching English in China, and by then I'd begun to grasp the various ways my students were often direct, sometimes blunt. It cropped up in their journals, where I'd asked them to record their daily thoughts about anything that interested them. Usually it was about their classes or friends, but sometimes I appeared, in slight caricature. My nose was large, skin dotted with freckles and too tan. I spoke too loudly and laughed too much with the other foreign teachers between classes. And now I was fat.

By then I'd learned that my students weren't being rude; they were making observations of the world in front of them – just like I'd asked. But it certainly wasn't what I'd expected. And that's how it crept up on me: a lingering case of culture shock.

During my first weeks of living in China, bumbling through the language and sorting out the basics like how to feed myself, every day was an adventure. A mall near campus had an entire floor comprised of over a dozen tiny restaurants, and I tasted my way through China – the spices of Sìchuān and Húnán food, Guǎngdōng-style seafood or local Jiāngsū's sweet-and-sour dishes. I made friends with the staff, deciphering conversations with basic Mandarin and a dictionary, learning where they were from and how many kids they had.

A few days after Tina's diagnosis, I wandered into the campus printing office, a tiny space in the administration building with two archaic laser printers. The only person on staff was a slight, birdlike woman named Miss Xu who darted between the printer trays and cubbies that lined the wall. I held up a worksheet for my students and said '60', not really knowing the Mandarin word for 'copies'.

She understood this request and snatched the paper from my hands. After a few moments she returned with a stack of warm photocopies. Then she said something I couldn't understand. I smiled and said 'Xiexie' – thanks – started moving towards the door. She repeated this

mysterious sentence. I uttered a vague 'uh-huh', still moving towards the door. She approached, repeating key words of the original sentence, this time in a slightly different order. 'Qian' I thought I heard. Money.

Of course! I needed to pay for the copies. I produced a wad of cash, but she waved it away and repeated her statement. 'Ting budong,' I said. I don't understand. She waved her hand again, this time shooing me away, the meaning of whatever she was trying to tell me lost in the silence between us.

In time, I learned to navigate moments like this with a healthy dose of humility, sometimes enlisting the help of Daniel, one of my students whose English was far better than my Mandarin. I learned, for example, that printing costs are allocated to your department, and Miss Xu was asking me what department I belonged to.

In the years of travel that followed my experiences in China, I linger on moments like the ones with Tina and Miss Xu, savouring the miscommunications and moments lost in translation, of getting lost and finding your way again. I let these experiences inform who I am as a traveller because I know that culture shock is just an unfamiliar world becoming a little more familiar.

"By then I'd learned that my students weren't being rude; they were making observations of the world in front of them – just like I'd asked."

Above (left and right): School children in Běijīng; Amish farmer on a horse-drawn plough.
Previous page: Chinese girl in traditional dress.

HAGGLE IN THE SOUKS
Cairo, Egypt
Filled with rich scents, colourful sights and frenetic noise, the souk (market) is where daily life unfolds, and Khan Al Khalili is Cairo's largest. Haggling is a time-honoured tradition, a way of socialising and meeting new people. High-value items like gold or antiques are common items where haggling is expected. Keep it friendly, though; it's a game – not a battle.
HOW The souk isn't close to any major public transport stops, so hire a taxi.

VISIT AN AMISH COMMUNITY
Pennsylvania, USA
The Amish are a group of traditionalist Christians known for their low-tech, simple way of living. Access to radio, TV and telephones are restricted, allowing these communities to maintain a centuries-old lifestyle. Although the Amish are a reserved bunch, some communities make their living introducing visitors to the Amish way of life. Take a horse and buggy ride, eat at an Amish-style restaurant, or just relax and soak in the charming landscapes.
HOW The Amish Experience (www.amishexperience.com) offers tours to Amish farms and homes in Lancaster, Pennsylvania.

CHOW DOWN ON INSECTS
Bangkok, Thailand
Eating insects might not be everyone's idea of fun, but in Thailand bugs such as grasshoppers, water bugs, silk worms and crickets are considered a savoury snack. Many of Bangkok's popular streets, including Khao San Rd, have street vendors peddling the cooked critters from wheeled push carts. Crickets and grasshoppers are typically more appetising, having a crispy, salty flavour.
HOW The insect vendors on Khao San Rd usually come out at night.

Witness the launch of a rocket to see the raw, awesome power of what human beings can do when they put their minds to it.

SEE A RO

Since the very beginning of time, humans have looked up at the sky in wonder, eager to explore its vast mysteries, as if beyond it are the answers to humankind's greatest questions: Where did we come from? Why are we here? Are we alone? Few of us earthlings are equipped with the right stuff to actually visit the stars, but seeing a rocket launch is the next best thing – from the early *Saturn V* rocket, as tall as a skyscraper, to the modern-day *Falcon Heavy*, which might even be boldly blasting off to new heights as you read this. Being present for a launch is to witness humanity's universal characteristics in one exciting moment – curiosity about the universe,

a hopeful view of the future and a unifying desire to exceed the limits of our imagination.

Prior to the launch, anticipation runs high. In the crowd, people are giddy at what they are about to experience. The countdown begins: 10, 9, 8... It rings through like a chant. 3, 2, 1... A distant hiss of igniting engines. Then lift-off, and the rocket is arcing towards the sky. It only takes a few minutes for the craft to be lost in the great blue above. But perhaps the best aspect of seeing a rocket launch is the people around you. Even now, half a century into the Space Age, the audience that gathers to watch a rocket launch will look at each other as if to ask: Can you believe it?

CKET LAUNCH

TAKE IT FURTHER

Marvel at the wonders of the world:
Behold a travel icon, p212
Reach a limit:
Stand at the end of the world, p234

THE SPACE COAST
Florida, USA
Spacecraft spectators have been flocking to a strip of coastline in eastern Florida since the early days of the American space programme. Today, the so-called Space Coast is home to a lift-off several times a year – either from the Kennedy Space Center or Cape Canaveral Air Force Station. In between launches, spaceheads can get their astral fix by visiting the KSC Visitor Complex to learn about the history of space exploration – and its future, of course.
HOW See www.kennedyspacecenter.com for the launch schedule.

BAIKONUR COSMODROME
Kazakhstan
Although the USA is often said to have won the Space Race by planting a flag on the moon, nearly every prior achievement was by the Soviets. That lives on at Baikonur Cosmodrome, a spaceport in southern Kazakhstan that's leased to Russia. Among the spacecraft launched here are the venerable Soyuz rockets, in operation since 1966.
HOW Baikonor Cosmodrome Tours (www.baikonurtour.com) offers five-day/four-night tour packages, which include flights from Moscow, accommodation and an English-speaking guide.

WALLOPS FLIGHT FACILITY
Virginia, USA
One of the world's oldest launch sites, Wallops Flight Facility was established in 1945. The spaceport now hosts six separate launchpads, sending large-scale rockets such as the 139ft (42m) *Antares* into orbit over Virginia's eastern coastline. The facility also sends up 'sounding rockets', or research craft, which carry measurement and experimentation systems and equipment into orbit.
HOW Sounding rockets and large-scale rockets can be viewed from the NASA Visitor Center on Wallops Island. See www.esvatourism.org for the launch schedule.

Left: NASA Space Shuttle launch in the USA

SEEK OUT THE RELICS OF AN ANCIENT CIVILISATION

Whether it's an arrowhead, a pottery shard or a giant pyramid rising out of the jungle in front of you, there's wonder in realising that other civilisations have prospered long before ours.

BUILD UP TO IT

Feel respectful:
Seek out sacred places, p46
Never forget:
Learn about the darker side of history, p86

It's easy to forget that we are a mere link in a long chain of human development that goes back, literally, to prehistoric times. There may not have been Twitter or email, but vast civilisations thrived in places that have since been completely deserted, abandoned except by archaeologists who painstakingly unravel the mysteries of what went on centuries ago. Climbing a pyramid, looking into the abode of a cliff-dweller or holding a 10,000-year-old pot in your hands connects you to these ancestors, working as a visceral reminder of not just how far we've come but how universal certain aspects of our human existence are.

Seeing it at first hand, looking up at pyramids or pausing to ponder the grief a mother must have had laying her baby's body to rest, the care with which someone carved a gift for his lover into stone — this understanding helps us focus on what's truly important, to escape the traps of modernity and to ground ourselves in the universal truths that we share. Leaving, you can't help but feel the journey wasn't just to a place but to a time.

LA CIUDAD PERDIDA
Colombia
Hidden deep in the Sierra Nevada de Santa Marta mountains of northern Colombia, this lost city was originally built and occupied by Tayrona Indians between the 8th and 14th centuries and is believed to be one of the largest pre-Columbian settlements in the Americas. The four- or five-day return trek, through steamy jungle and across rivers, is thrilling.
HOW Book a guided hke in Santa Marta; visiting the site alone is prohibited. Turcol (www.turcoltravel.com) are the most experienced.

Left: The Great Sphinx and pyramids at Giza

UENOHARA JŌMON SITE
Japan
Discovered in 1997 at a Kagoshima building site, this is one of the oldest sites of human habitation in Japan. Jōmon culture existed from about 14,000 to 300 BC. Next to the archaeological site is a museum, and it's possible to walk through reconstructed *hogans* (huts) and see what the hunter-gatherer culture was like at the time, as well as learn about their connections with modern Japan.
HOW Take a train to Kokubu station in Kirishima City. From there, Uenohara's Jōmon-no-Mori museum (www.jomon-no-mori.jp) is about 30 minutes by bus or 20 minutes by taxi.

ANCIENT PUEBLOANS
Mexico
Once called the Anasazi, this culture and variants of it spread from Utah deep into the northern part of modern Mexico. Whereas some areas, such as Mesa Verde and Chaco Canyon, are preserved in national parks, numerous other ruins lie along riverbeds and ravines, unattended relics of a vast and vanished civilisation.
HOW Many of the cliff dwellings in this area require lengthy treks. Recapture Lodge (www.recapturelodge. com) in Bluff, Utah, is an excellent base for exploring. Renting llamas (www. llamapack.com) is an option if hiking in the canyons.

GIZA
Egypt
Nothing compares to the structures rising out of the desert sands, visible for miles. Iconic symbols of all the mysteries of the past, the Sphinx and pyramids at Giza remain the zenith of everything lost civilisations hold. From secret chambers filled with treasure, to the curses of mummies, to conspiracy theories and links with alien visits, these giant monuments resonate deeply with anyone lucky enough to visit them.
HOW Cairo-based companies such as DJED Egypt (www. djedegypt.com) offer flexible tours that can include the pyramids of Giza, a Nile cruise, Luxor and other sites.

GET OUT OF YOUR COMFORT ZONE

BEGINNER'S MOUNTAINEERING

Fontainebleau, France
Too scared to climb mountains? Start with a little light bouldering at Fontainebleau, 30 miles (50km) south of Paris. Considered one of the world's best bouldering spots, the forest here is scattered with thousands of sandstone boulders of all shapes and sizes, for all levels of climber. The perfect spot for a climbing apprenticeship.
HOW Read the guide at www.boulderfont.info.

RIDE HIGH COUNTRY

Victoria, Australia
Discover Australia's Alps on horseback, with breathtaking views of the Great Dividing Range running 2175 miles (3500km). High Country rides start 5000ft (1500m) above sea level on the Bennison Plains at a mountain hut, and follow the scenery of classic Aussie film *The Man from Snowy River*.
HOW Rides cover three–seven days; droving trips are also offered. www.snowyrivertours.com.

JOIN THE CIRCUS

New York City, USA
Throwing it all in, running away and joining the circus is the classic life-change fantasy, and anyone can try it on for size in New York City. At the Circus Warehouse you can take classes in circus arts, from flying trapeze to contortion to wire walking and everything in between.
HOW The Circus Warehouse (www.circuswarehouse.com) offers specific classes and intensive training programmes.

TRY COASTEERING

Cornwall, England
This amphibian pursuit lets you get intimate with the coastline while surmounting a series of land- and water-based physical challenges. Clinging to the cliffs and ledges edging the sea, you'll shuffle and slide over rock stacks, belly-flop into plunge pools and splash through waves.
HOW The Newquay Activity Centre (www.newquayactivitycentre.co.uk) offers Cornish coasteering adventures.

ICY TIMES

Finland
Slipping into an ice hole in a lake of freezing black water is not, for most of us, a tempting prospect. But in Finland, where winter lows get down to -30°C (-22°F), many people start every day with an ice swim to benefit from the resulting energy boost. You can't knock it till you've tried it.
HOW Experience it in Rovaniemi in the country's north, in the chilly months between November and March. www.visitfinland.com.

TAKE IT FURTHER
Be moved by music:
Go out dancing, p62
Enjoy the benefits of a
cold-water swim:
**Take a plunge in
open water, p96**

That's when things start to get really interesting – when we push ourselves beyond what we know we can do. Prod yourself to reach beyond your known boundaries: you never know what you'll find there.

Above: An ice swimming hole on a lake in Kuhmo, Finland

DUNE BASHING
Qatar
The ultimate adrenaline rush or heart-stopping death ride? Dune bashing, as practised on the enormous sand dunes of Qatar, requires a large SUV, equipped with a roll cage in case of overturn. If you're self-driving, carry a tow chain and watch out for quicksand; better still, take a tour.
HOW Inbound Tours Qatar (www.inboundtoursqatar.com) offers sand dune bashing trips out of Doha.

GO ON SNAKE SAFARI
Kenya
The Big Five are yesterday's news. Take a real African wildlife challenge and hunt down native reptiles around Tsavo National Park. Of Kenya's 127 snake species, only 18 have caused human fatalities, so the risks are mitigated – but it's best to be in expert hands when you confront a puff adder or rock python in the wild.
HOW Bio-Ken Snake Farm (www.bio-ken.com) runs safaris and tours.

SANDMINE CHALLENGE
USA
SpeRUNking, as opposed to spelunking, involves running through the cave. Or crawling, wading, dodging obstacles and quicksand, as necessary. Its world championship is held in Missouri's Crystal City sandmine, 3 miles (5km) in length, with sand dunes and an underground lake.
HOW Visit Sandmine Challenge (www.sandminechallenge.com) to register.

GO ROLLING
New Zealand
Rotorua is the home of zorbing (aka globe-riding, sphereing or orbing). The gentle hills on the side of Mt Ngongotaha, overlooking Lake Rotorua, are an ideal spot to roll around in a giant inflatable ball. (It's actually two balls, one inside the other, so you're cushioned from rocks and bumps.)
HOW You can Zorb at OGO Rotorua (www.ogo.co.nz) or Zorb Rotorua (www.zorb.com).

CRATER-GAZE
Kamchatka, Russia
Kamchatka is an ancient version of earth, a primal wilderness populated by bears, moose and wolves, and crowded with about 300 volcanoes. Thirty are still active, so you're guaranteed to experience one, maybe even two or three.
HOW Volcano Adventures (www.volcanoadventures.com) organises a three-week expedition in Kamchatka.

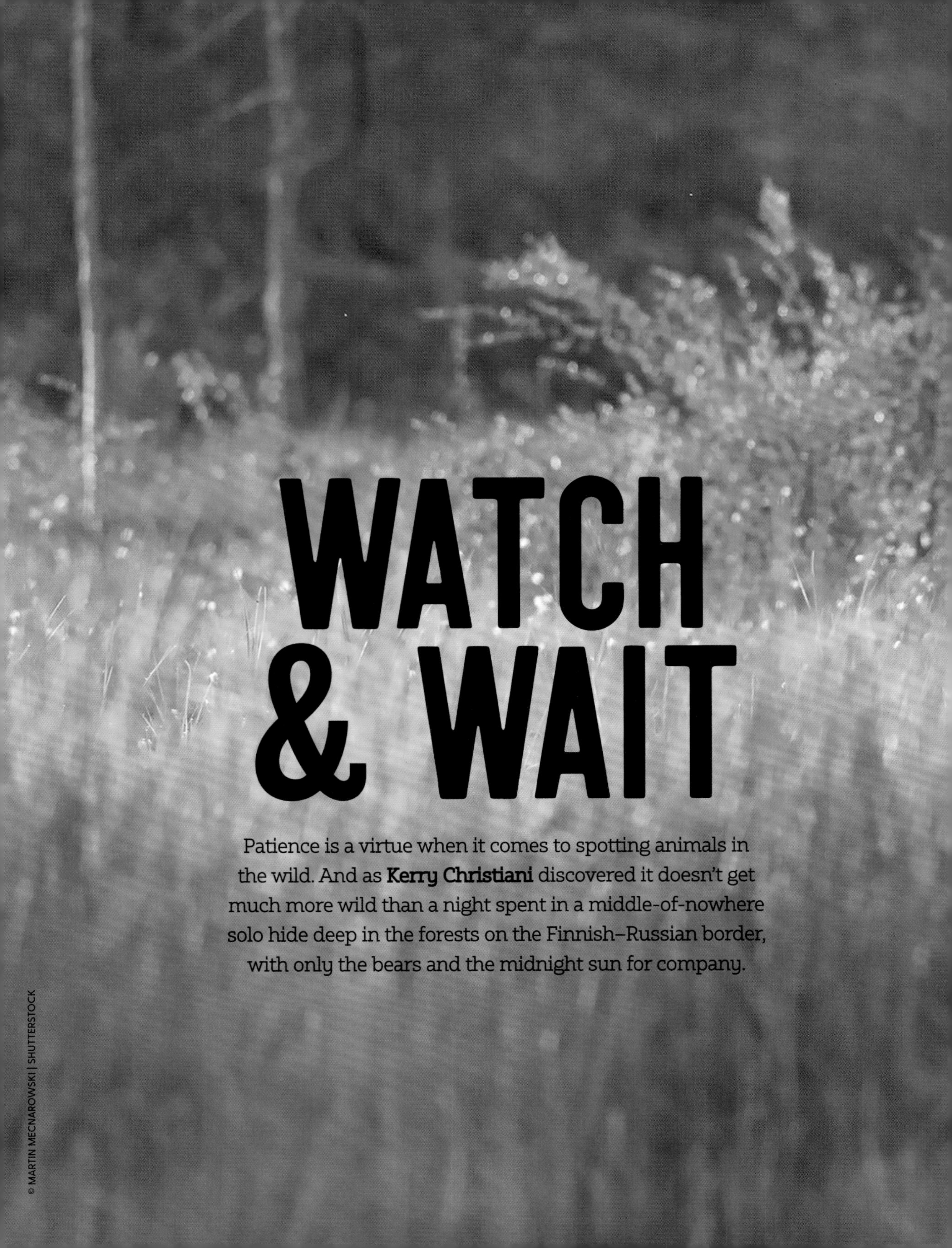

WATCH & WAIT

Patience is a virtue when it comes to spotting animals in the wild. And as **Kerry Christiani** discovered it doesn't get much more wild than a night spent in a middle-of-nowhere solo hide deep in the forests on the Finnish–Russian border, with only the bears and the midnight sun for company.

© MARTIN MECNAROWSKI | SHUTTERSTOCK

It's around 1.30am and I'm just starting to drift off in the twilight of midsummer when I hear the low grunt and hefty squelch of a brown bear plodding through the swamps towards my solo hide. In the befuddled daze of half sleep, I think I might be dreaming. No, there it is again. It's definitely a bear: a walloping great bear – and it is close, it's really close. I feel as though my heart will burst at any second, and I am frozen with fear, my palms clammy with cold sweat. How near is it? A metre away? Maybe two? Near enough for me to hear its rasping breath and hungry snuffles, that's for sure. It would help if I could see the bear, but I'm curled up in an awkward ball on the floor. Don't move a muscle, I think, as I try to control my quickening breath.

Logic eventually kicks in long enough to recall the words of Sabrina, a guide at the Wild Brown Bear Centre in Kuhmo, Eastern Finland, who has spent 10 years in the bear-watching business. Earlier that evening she had led me through swamps of pearl-tipped cotton-grass and spruce forest swarming with bloodthirsty mosquitoes and thick with lingonberries to a rickety shack that a bear could almost certainly turn over with a single paw. 'Don't worry,' Sabrina had said, reassuringly. 'The brown bears in Finland are quite friendly – not like their grizzly cousins in North America. They hear you, they smell you, but they can't really see you. They have poor eyesight.'

Staying the night in a solo hide had seemed like a fun idea at the time, but suddenly I was faced with the reality. 'That's where you pee,' Sabrina said, pointing to a hole in the corner. 'That's where you can sleep, if the bears don't keep you awake,' she said, nodding at a dusty square metre of floor space. 'And this is where you can take photos of the bears,' she said, indicating a tiny window without so much as a pane of glass. With nothing but a flimsy curtain between me and one of the world's greatest predators, this was going to be quite some night. Sabrina noticed the concern crossing my face. 'Ah, you'll be fine,' she said. 'But whatever you do don't leave the hide until I return at 6am tomorrow.'

The light faded and heavy-lidded monotony began to set in. Hour after hour passed and it appeared to be a no-show. I squinted at the dark pencil line of forest, trying to decipher if the shadows were bears or rocks, watching the light grow thinner. Then, almost when I had given up hope, it happened: a brown bear swaggered from the thicket, rising up on all fours to survey its 'hood.

As if word about picnic time had spread, another bear traipsed around the lake in my line of vision, grumbling in search of a midnight snack: lingonberries, insects, roots, whatever was going. As I clicked away in stunned silence, a wolverine emerged stealthily from the gloom – a rarity even in Finland's back of beyond – then disappeared almost as quickly as it had appeared.

But now, having given in to tiredness, I'm crouched on the dusty floor of the hide, with my pulse racing and pure terror flowing through my veins. It's as much as I can do to haul myself upright and inch towards the window, slowly, slowly, ever so quietly. The bear seems to momentarily meet my gaze, despite its supposedly faltering eyesight. Then, with a final contented grunt, it turns its back on the bounty of Finland to wander back over to Russia, back to the woods, back to bed.

HOW Adventure operator Explore (www.explore.co.uk) offers a three-night brown bear-watching weekend at the Wild Brown Bear Centre in Kuhmo during the peak bear-watching season (late May to mid-September), including guides, meals and local transport. You can also make your way to the centre independently to stay the night in a bear photography hide. Before booking wildlife-related activities, check the company's environmental credentials carefully. Look for conservation-focused operators that do not threaten or scare wildlife.

"With nothing but a flimsy curtain between me and one of the world's greatest predators, this was going to be quite some night."

Above (left and right): Brown bear looking out from the forest; mountain gorilla, Uganda.
Previous page: Brown bear

MOUNTAIN GORILLAS
Uganda

Uganda has paved the way when it comes to the conservation of mountain gorillas, with more than 450 of the endangered apes and counting. Bwindi Forest National Park has the world's highest concentration of the gentle primates. Tackle the stiff hike into the jungly mountains and you'll see gorillas feeding, resting and playing in their natural environment.
HOW Year-round guided tours abound but permits are in high demand (book at least six months ahead). A reputable operator like Yellow Zebra Safaris (www. yellowzebrasafaris.com) can take care of the logistics.

MUSK OX
Norway

Thanks to its Arctic-like habitat, Norway's dramatic, mountainous and delightfully remote Dovrefjell-Sunndalsfjella National Park is one of the few places in the world where you can witness mighty musk oxen. Go on a guided hike through beautiful alpine landscapes for up-close encounters with these hefty, woolly beasts. Reindeer, mountain grouse and large birds of prey can also be sighted.
HOW Oppdal Safari (www.moskussafari.no) runs full-day musk oxen safaris during summer (June to September), departing from Oppdal station.

JAGUARS
Brazil

The immense, biodiverse wetlands of the Pantanal in southwestern Brazil offer the chance to observe jaguars in the wild. Prime time is the dry season (June to October), when the jaguars are forced to head for the riverbanks. Even so much as a fleeting glimpse of the world's third-largest cat at close range is a heart-pounding experience.
HOW Pantanal Jaguar Safaris (www. pantanaljaguarsafaris.com) offers a multiday Porto Jofre Jaguar safari. Prices include lodge accommodation, guides, excursions, transfers and meals. The gateway airport is in Cuiabá.

TAKE A VINTAGE TRANSPORT METHOD

Sometimes it's not where you're heading but how you're planning on getting there that causes the most excitement, especially if the journey involves a form of transport that makes it feel as though you're travelling back in time.

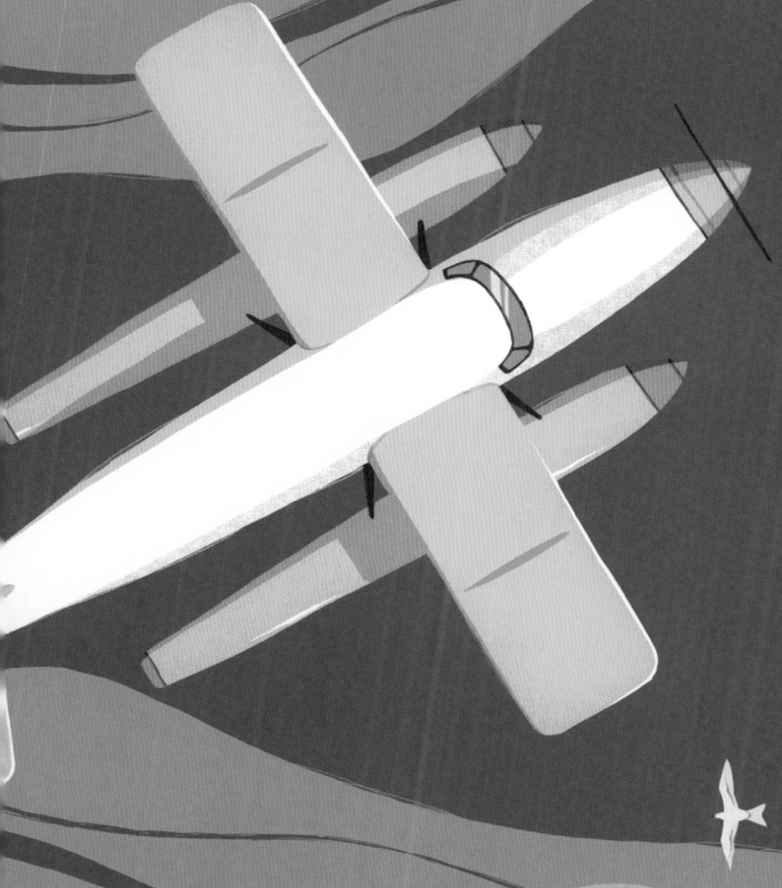

CLASSIC BIKE THROUGH THE HIMALAYA

India

Take a tour through the world's highest passes astride a classic Royal Enfield Himalayan motorbike. Named after the mountains, and now made in India, these beautiful booming beasts are ideal steeds for wending past waterfalls, snowmelt rivers and the occasional yak.
HOW The Himalayan Heights tour covers 870 miles (1400km) over 13 days, reaching 17,650ft (5380m). See www.rideexpeditions.com for more.

FLOATPLANE

Fiji

Recreate a time when flying boats on the Coral Route skimmed across the South Pacific – bouncing between the islands and lagoons of Fiji, Samoa, Tonga and the Cooks – by taking a floatplane from Nadi to the Yasawa and Mamanuca Islands.
HOW Flights to the Yasawa and Mamanuca islands operate daily; others are by charter booking. See www.pacificislandair.com.

DARJEELING RAILWAY

India

This 140-year-old railway sees British-built B-class steam and diesel locomotives travelling along narrow-gauge tracks past tea plantations, villages and mountain vistas.
HOW The steam service travels between Darjeeling and Ghum, India's highest station. See www.dhr.in.net for more.

PADDLE STEAMER

UK

Board the world's last sea-faring paddle steamer, the PS *Waverley*, and take a time-bending voyage through the Western Isles or join a leg rounding Land's End.
HOW Full timetables are available at www.waverleyexcursions.co.uk.

TAKE IT FURTHER

Go on your own steam:
Take a big trip alone, p240
Cross continents in one long journey:
Make an epic overland journey, p268

© DAVID DORAN

ENGAGE ALL

Your senses go on a binge when you hit the road, stoked by an endless array of multihued, salty-sweet, stinky-perfumed, hot-cold, car-honking, bird-chirping experiences.

YOUR SENSES

Sight is the usual sense that gets deluged when you travel, and indeed, it's the one your brain gives preference to for processing information. You usually see your destination first: the fairytale castles in Germany, the red-rock centre in Australia, the prowling lions in Botswana.

Then the other senses kick in. Taste is a favourite to indulge, from salty yak butter tea in Tibet to sweet hazelnut gelato in Italy to searing sambal chilli sauce in Indonesia. Smell is the sense most linked to memory, which is why the scent of orange blossoms reminds you of your time in Seville, the whiff of yeasty bread morphs you to Paris, and the aroma of diesel fumes sends your brain back to Kathmandu's tumultuous streets. Sound adds to the sensory jam. Think chittering monkeys in Mexico's Yucatán jungle, clattering pans at Hanoi's street stalls and salsa music blasting from Cuban radios. As for touch, it's perhaps the most underrated sense when travelling, but anyone who has ever felt the soft sand on a Maldives beach or the steamy water in an Icelandic hot spring knows its value.

The key is to fire up all five senses at the same time, which heightens your experience and leads to a more encompassing sense of place. Try it next time you're out: make an effort not just to see what's around you, but to hear, smell, taste and feel it, too.

TAKE IT FURTHER

Show up and appreciate someone:
Meet your hero, p254
Be empowered:
Face your fears, p266

Left: Varanasi and the banks of the Ganges river

VARANASI
India

India's holiest city explodes your senses. There's the smell of cow dung and fires burning on the ghats where bodies are cremated. Tabla drumming and twangy sitar music waft from open windows. But it's what you see that strikes you most: loin-clothed sages, flower sellers, women in jewel-bespeckled saris and everyone else – the rich and poor, the water buffaloes – bathing in the sacred Ganges River.
HOW Varanasi Walks (www.varanasiwalks.com) offers 2.5-hour city tours by foot.

SALVADOR
Brazil

Afro-Brazilian Salvador pops the eyeballs with its brilliantly hued colonial-era buildings and gold-laden churches. Samba-reggae music blasts your eardrums, while the smell of street vendors sizzling fritters floats by your nose. And coconut, coriander, hot peppers and fish stew ignite your taste buds, thanks to the city's classic dishes.
HOW Head to the Pelourinho district for the sensory jackpot. Restaurante do Senac (www.ba.senac.br) is perfect for a tempting buffet of regional dishes.

ZANZIBAR TOWN
Tanzania

Wander the maze-like streets in the historic quarter and you'll see something new at every turn: a former palace, a Persian bathhouse or a coral-stone mansion with latticework balconies. You'll hear the muezzin's unmistakeable sing-song call to prayer, smell jasmine trees and cloves, and feel the warm waves of the Indian Ocean.
HOW Zanzibar Different (www.zanzibardifferent.com) runs tours that go to spice farms, markets and local homes for a meal.

LIVE ON THE FRINGE

Exploring a minority community in a foreign land can give you a valuable new perspective of a place – often one that is more complex, and accompanied by stories of hope, struggle and what it means to call a place 'home'.

Many of the most interesting places owe their vitality to multiculturalism. On London's Brick Lane, thriving Bangladeshi restaurants are neighbour to art galleries and fashion stores, while in Paris, North African cuisine and music are woven into the fabric of the streets. People may move away from their countries – seeking a new life or better working opportunities – but native cultures, cuisines and language are harder to leave behind. And where one family has made a home for itself, others will follow – through word of mouth, while courage is still high. And then the first ethnic shops and restaurants start popping up like bright beacons of another home. These minority communities exist all over the world, giving insight into the struggles and successes of immigration. America's beloved Statue of Liberty used to greet newcomers to New York, now one of the most dynamic cities in the world thanks to hundreds of foods, languages, and cultures intermingling as one. When you travel, you're probably envisioning a destination you've predetermined to some extent – people will look like this or speak this language. But if you search a little beneath the surface, you'll find a rich tapestry of voices combining to create the essential fabric of a place, and you might discover a view into two cultures: the minority *and* the mainstream.

CHINESE COMMUNITY IN JOHANNESBURG
South Africa

Johannesburg has the largest population of Chinese immigrants on the African continent. These days, many move there in the hope of starting small businesses.

HOW Wander along Derrick Ave, the heart of Chinatown, then head north to Chinese Northern Foods Restaurant on Rivonia Blvd for the best handmade dumplings in the city.

CUBAN COMMUNITY IN MIAMI
USA

During the Cuban Revolution of the 1950s, hundreds of thousands of refugees moved across the ocean to the USA. It's impossible to visit Miami without experiencing its Cuban influence, especially in the food and music.

HOW The jewel of the Little Havana Art District, Cubaocho is renowned for its concerts, as well as being an art gallery and community centre; www.cubaocho.com.

Below (left and right): Street scenes in Little Havana, Miami

GERMAN COMMUNITY IN BLUMENAU
Brazil

Located in southern Brazil, Blumenau shows off its Bavarian heritage, especially in October when it is host to the second-largest beer festival in the world.

HOW Oktoberfest runs for three weeks in October. Alternatively add Bier Vila to your itinerary. Thanks to 30 options on draft and 300 local and regional craft beers by the bottle, it is Blumenau's best drinking den for enthusiasts.

TIBETAN COMMUNITY IN CHENGDU
China

Many Tibetans relocate to Chengdu for its proximity to home and better working opportunities. The Tibetan Quarter, just south of Wuhou Temple, is dubbed Little Lhasa.

HOW Book a bamboo bicycle tour through the district with Extravagant Yak; www.extravagantyak.com.

© TORONTONIAN | ALAMY STOCK PHOTO, © JUSTIN FOULKES | LONELY PLANET

Staying with Syrian refugees in Turkey

"Ghazwan was hosting three couch-surfers besides myself: a German on tour around the world and two stocky Russian best friends. I'll never forget what a motley crew we must have looked like, tramping through the streets of the Fatih District, or 'Little Syria', in Istanbul together. Nonetheless, we were greeted at the door of his favourite restaurant with open smiles.

As we ate our hummus and roast chicken, Ghazwan spoke candidly of his difficulties adapting to his new country. He was an English professor who'd fled Syria in 2011. He considered himself well settled – enough to rent a tidy apartment and to sponsor his nephew's visa – but he still had strong pangs of homesickness. He felt particular despair, he said, thinking of the war in his homeland. But he was happy here.

Staying with Ghazwan was an eye-opening experience for me, someone who also comes from an immigrant family. I found myself re-examining what it meant to be 'home', or to discover 'home' in new places, with new people. After we finished eating, we stepped out of the restaurant and into a cloud of scent from the self-service laundry and perfume shops next door. I tilted my head back, sure that there had never been anything as lovely as the sounds of Arabic and Turkish mixing, floating upwards from the street to the sky."

Rucy Cui, publicity associate

TAKE IT FURTHER
Slow down and focus on a new skill:
Learn a craft, p94
Enjoy sustainable community living:
Go off grid, p192

Get lost! Well, not literally. It's a good idea to know vaguely where you are, but escaping the modern world and its reliance on fossil fuels and connectivity has surprising rewards. After all, the worst mobile phone service often comes with the best views, and where convenience (and reliable electricity) is lacking, nature might just blow your mind.

*L*iving off the grid for a while isn't just about taking a digital detox, though you probably need it. It's about re-rooting yourself in nature, and remembering how it feels to walk with a smaller footprint, leaving no trace behind. It's about sustainable living. Scratch that, it's about sustainable living in a drop-dead gorgeous location.

All around the world – from the Andes to Alaska, and from Hawaii to the Himalayas – eco-camps and low-impact resorts have sprouted up to take advantage of nature's wonder without simultaneously draining its resources. Here you will eat whatever is locally available; keep warm from the power of renewable energy; and live and sleep by the natural rhythms of the day.

It's all part of living in harmony with the environment, and when the living is this good, you'll soon be noting tips and ideas as to how to continue the way of life at home. Because even in suburbs and cities, it's possible to inch your way off the grid, we can all eat more mindfully (maybe even grow our own), compost, and pedal instead of drive, and generally reduce our overall impact. Imagine if everyone did that. What a world we might create together. And, yes, fine, you can take pictures with your smart phone (and solar charge it too), so you can always remember the trip that changed your life.

GO OFF GRID

Take it further
Remove distractions and spend quality time with yourself:
Disappear, p262
Learn to co-exist with nature:
Live off the land, p272

MOUNTAIN LIVING
Chile

In the remote reaches of Torres del Paine National Park, Patagonia, two Chilean engineers have created the fully sustainable EcoCamp Patagonia. Inspired by the leave-no-trace lifestyle of the indigenous Kaweskar people, guests stay in custom-built geodesic domes, which are studded with skylights, and powered by solar and hydraulic energy.
HOW Packages include dome stays with trekking and wildlife tours; www. ecocamp.travel; go in summer (Nov–Mar).

SURFSIDE SOUL
Mauritius

In a prime beach-side position on the East African island nation of Mauritius is a slice of sustainable, multicultural paradise. What makes Salt Palmar special isn't just the gorgeous location, activities, or sumptuous spa, it's the fact that everything is designed to minimise impact without compromising on luxury. Food is organic and grown locally, fish is sourced sustainably, and they are vegan-friendly and plastic-free.
HOW Go in the dry months (Jul–Aug); www.saltresorts. com.

SOLAR-POWERED HIDEAWAY
Guatemala

On its own beach overlooking stunning Lake Atitlán, and with panoramic volcano views, sits La Fortuna at Atitlán. Guests stay in private, palm-roof cabañas built from wood and glass. It is all solar-powered, except for the hot tub, which is heated with a wood fire. The food is tasty and prepared from locally sourced produce, and the bar is sensational. No wonder it was named one of the top 25 small hotels in the world in 2017.
HOW Dec–Jan are the best months to visit; www. lafortunaatitlan.com.

Left: Sun setting over Lake Atitlán, Guatemala

VISIT
EMPTY
PLACES

Svalbard is Europe's largest continuous wilderness and the final frontier before the North Pole. Here, around 78° north, **Kerry Christiani** discovers a land shaped by mighty glaciers and defined by the coming and going of the sun.

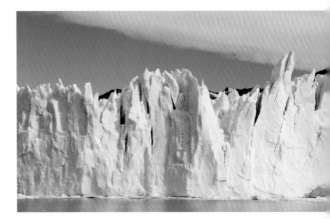

ilence is the sound of Svalbard, the archipelago dangling off Europe's northernmost fringes, and a mere 620 miles (1000km) of pack ice away from the North Pole. It is not that there is no noise: listen carefully on calm days and you might hear the sea ice shift, a husky howl, reindeer snuffling for food, the crunch of fresh snow beneath your feet. But the meaning of silence here is more profound than that. Beyond the main settlement of Longyearbyen, you enter one of the world's last empty wildernesses and the realm of the polar bears.

The light here has a special quality, too – there is either too much of it (in midsummer) or too little (during the long, long Polar Night). When I arrive in late February, the sun is glaring defiantly on the horizon after four months of absence, bringing with it unexpected colours. When I'd envisaged the Arctic, I'd pictured a bleak white wasteland. But now, as I bounce giddily through a gully, over frozen tundra and into glacial valleys on a snowmobile, this is what strikes me the most: the soft pinks and pale blues, the violets and lilacs. These are the cut-glass colours that you only get in bone-chilling cold. It's around -20°C (-4°F) and an icy wind is whipping my visor and stinging and numbing my face. But oh, the beauty!

A five-day snowmobile expedition into the backcountry of Spitsbergen opens up a High Arctic of Narnia-like enchantment. Once you've forgotten the cold and mastered the knack of navigating the icescapes, you can start appreciating the details: the bare, muscular mountains seemingly rippling into infinity and glowing pearlescent in the blue air as if lit from within, the echoing silence as soon as the engines are switched off, the delicate patterns created by different ice formations, the Northern Lights flickering in night skies, the wildlife.

Near Kapp Linné on the west coast, as we rumble towards the blue glitter of Grønfjorden, I notice a speck in the distance. It turns out to be an Arctic fox moving swiftly along the fjord shoreline, barely noticeable in the snow because of its pure-white winter coat. 'The best bit about going on an expedition is the wildlife you get to see,' says our expedition guide Marte Myskja. 'Reindeer, whales, sometimes polar bears.' Indeed, Svalbard keeps giving on this front. At Sassenfjorden, we make out a walrus with her two cubs, lolling on a

bergy bit. They seem insignificant from afar, but males are three times heavier than the average polar bear – and the more fearsome opponent of the two.

Huskies howl a greeting at the North Pole Camp – a sound that is truly of the north. As we arrive, the light is putting on a performance in its dying moments, with flashes of green illuminating the heated tents in which we will spend the night. Tripwires and flares are in place in case there's an unexpected visitor in the form of a polar bear. 'We're on the polar bear's territory, it's important we respect that,' says Marte. Later, as I tiptoe cautiously out to the toilet in the middle of the night, I can't help but look over my shoulder.

The remote isolation of Svalbard has a powerful effect on the psyche. This is a place still immune to time and trends, where polar bears are often out of sight, but never out of mind. As we cruise back to Longyearbyen, past frozen waterfalls and wide snowy valleys where wild Svalbard reindeer roam, I start to brace myself for the return to civilisation. We've only been gone a few days, but this feels like a kind of eternity in the Arctic.

HOW Spitsbergen is the major hub for striking out into Svalbard's wilderness; the main airport is located here. There are frequent direct flights to/from Oslo to Longyearbyen on Spitsbergen with Norwegian Airlines. The journey takes three hours. Sandgrouse Travel (www.sandgrousetravel.com) offers a five-day, four-night trip. Basecamp Explorer (www.basecampexplorer.com) in Longyearbyen is designed like modern-rustic trapper's lodge and has a cosy lounge for post-expedition chilling.

> *"The remote isolation of Svalbard has a powerful effect on the psyche. The place is immune to time and trends"*

Above (left and right): Parque Nacional Los Glaciares in Patagonia; driving in the Empty Quarter, UAE. Previous page: Deserted valley near Longyearbyen, Svalbard

EMPTY QUARTER
Saudi Arabia, Oman, Yemen and UAE

Empty by name and nature, Rub al-Khali, otherwise known as the Empty Quarter, is the largest contiguous sand desert on the planet, rippling for some 250,000 sq miles (650,000 sq km) across Saudi Arabia, Oman, Yemen and UAE. This vast wilderness, which captured the imagination of Lawrence of Arabia, is one of the hottest and least hospitable places in the world. The best time to visit is from October to April.

HOW Take an expedition on camel or by 4WD with a night under the stars. A reputable operator is Bediyah Safari Tours, offering a two-day trip to the Empty Quarter (including transport, guides, meals and camping gear).

THE GOBI DESERT
China and Mongolia

Spreading across northern China and southern Mongolia, the Gobi means 'waterless place'. Epic sand dunes, dinosaur fossils, yurt-dwelling nomads, pencil-straight roads and wide-open horizons, bare rock, fierce winds and gold-green steppe – this desert has the lot. And the spangled night skies are something else. There are few better places in the world in which to disconnect.

HOW June and September are good months to visit. Stay with a nomadic family, cross the desert by 4WD or explore on horse- or camel-back. Wild Frontiers (www.wildfrontierstravel.com) offers a two-week camping and horse-riding expedition into the Mongolian wilds.

PATAGONIA
Chile and Argentina

Ferocious, ice-cold winds are often the first thing to hit you on arrival in Patagonia. At the southern tip of Chile and Argentina, this barren, harshly beautiful land beguiles with jewel-coloured lakes, crashing glaciers, dark Mordor-like mountains, and the steppe where guanacos roam and condors wheel in china-blue skies.

HOW The hiking season runs October to April (spring to autumn). Swoop Patagonia (www.swoop-patagonia.com) and Walk Patagonia (www.walk-trek.tur.ar) run multiday treks to remote corners of Parque Nacional Los Glaciares. Prices and dates are available on request and programmes can be tailored accordingly.

TAKE A BIG TRIP ALONE

LOOK INTO THE EYES OF A PREDATOR

GO MEAT-FREE ON THE ROAD

TRAVEL WITHOUT PLASTIC

MEDITATE WITH

ST YOUR METTLE

CHALLENGE YOUR PERCEPTIONS

TRAVERSE A COUNTRY ON FOOT

HELP SAVE AN ENDANGERED SPECIES

SURVIVE IN THE WILDERNESS

MASTERS

APPRECIATE THE ORDINARY

5

APPRECIATE THE ORDINARY

Fruit, vegetables, a soft bed, a cold drink – things that seem ho-hum at home can take on a new shine when you're journeying.

You've probably experienced it, especially if you've ever travelled outside your comfort zone: how something that's normally unremarkable transforms into something amazing and sought-after when you're on the road. Often it's a simple thing, and you don't realise how much you value it until it's no longer readily available. It might be sunshine if you're spending your winter in Tromsø, Norway, where darkness is constant from November to January. Or it might be elbow room if you're in Dhaka, Bangladesh, where people are jammed tight into the most densely populated city in the world. Or maybe

it's ice if you're longing for a cold drink and you're travelling in a warm-drink realm. The first time your mouth returns to cubes bobbing in a lemonade you almost pass out from pleasure.

A lot of this effect has to do with gratitude. Travel stokes it by teaching you to not take things for granted. That's why a soft bed after you've been on a week-long trek in New Zealand is perhaps the most rapturous thing you've ever felt, and a simple boiled pumpkin for dinner in a remote Indian village tastes Michelin-star sublime. Travel gives you the gift to find the magic in the mundane.

WATER IN THE NAMIB DESERT

Sesriem, Namibia

A walk among the 984ft (300m) dunes at Sossusvlei and the parched trees in the bleached-white ground at Deadvlei make you value what a dribble of river can do.

HOW Explore the landscape with Chameleon Safaris (www.chameleonsafaris.com) on a one- or two-night trip departing from Windhoek or Swakopmund.

FOOD ON THE PACIFIC CREST TRAIL

Western USA

There's nothing like a 2650-mile (4265km) trudge through desert and mountains to work up an appetite. Books have been devoted to the ravenous cravings that the trail inspires.

HOW Try the section through Yosemite, Kings Canyon and Sequoia national parks in California (www.pcta.org). Tuolumne Meadows Store in Yosemite is a popular pancake-and-ice-cream refuel stop.

Below (left and right): Thirsty terrain in the Namib desert; street food stand in Chóngqìng, China

QUIET IN CAIRO

Egypt

One of the world's loudest cities, Cairo throbs with the distorted roar of the muezzins' call to prayer, horns honking nonstop and beat-booming electronic music that blasts from nearly every shop. A quiet space is a precious commodity here.

HOW Find calm in the Garden City district at leafy Falak cafe and bookshop (www.facebook.com/falakbooksandartworks), open 9am to midnight daily.

BEING UNDERSTOOD IN CHÓNGQÌNG

China

This buzzing metropolis is one of China's largest cities. Yet it has few foreign visitors and isn't set up for those who don't speak Chinese. Sometimes all you want is a familiar tongue.

HOW English-speaking Suzie's Pizza is open 11am–10pm in the UME Building near Shapingba metro.

The enchanted turnip

"We arrived at the low stone buildings just as the sun shot below the mountains. Stillness reigned, except for the occasional flutter of a prayer flag. Our stomachs growled.

We were trekking north of Kathmandu and winging it off the beaten path, hoping that whatever village we arrived at each afternoon would have food and lodging. The plan wasn't working so well on this day. It was late and the place looked abandoned.

We sat down and started talking about the French fries and cream-filled doughnuts we'd eat if we ever reached civilisation again, when an elderly man in a maroon robe hobbled out of one of the buildings. He flashed three wiggly teeth when he smiled and waved us in. Turns out we were at a Tibetan Buddhist retreat for monks, unoccupied currently except for him.

His room had a bed, an altar topped with a silvery Buddha, and a fire pit. A pot bubbled over the latter. The monk threw in ramen noodles and … was that a giant turnip?

It was, and it was the most luscious meal ever. Tangy, salty, oniony. A turnip! Maybe we were starving and anything would have tasted good. Maybe he did something magical to it. But turnips have spellbound my taste buds ever since."

Karla Zimmerman

TAKE IT FURTHER

Ask more of yourself:
Challenge your perceptions, p210
Give thanks to inspiring people:
Meet your hero, p254

LOOK INTO THE E

Sometimes death can be staring you in the eye, a powerful, humbling reminder that there are creatures on this planet more powerful than we are.

S OF A PREDATOR

All too often we think of ourselves as the 'masters' of this planet. We go about our lives with the expectation that we'll be passing away peacefully in our sleep decades into the future. But many of us live in places where humans aren't the only apex predators, and it's very possible to find yourself in the wrong place at the wrong time. Whether it's on land or at sea, primordial enemies of humans still exist, and seeing them is an awesome, humbling reminder of our true place on earth.

Yet a journey to look into the eyes of a predator isn't just about fear. It's also about discovery, learning how endangered these incredible creatures are, due to hunting, destruction of their habitat or their active extermination. Yes, a cheetah could rip you apart, but it's in a desperate race for survival and facing great odds. Great white sharks are terrifying if you're swimming with them, but they are routinely hunted at fishing tournaments, often killed for someone's bragging rights. It's impossible to seek out these animals without also discovering how fragile they are and how much we need to protect them.

As you seek out these incredible experiences, use care and educate yourself on how to conduct them in a responsible way with the welfare of the animal always in mind.

BUILD UP TO IT

View the world from nature's standpoint:
Travel on horseback, p52
Commune with higher powers:
Meet the planet's giants, p136

KOMODO DRAGONS
Indonesia

Boat tours leave from Flores, Bali, and other Indonesian locations for trips to Komodo and Rinca islands, where you can take guided tours to see the world's largest lizard. Though you may see a dragon or two lazing around near the check-in booth, you'll want to journey into the jungle for a closer look. Heed the warnings from rangers and don't straggle; though attacks are rare, tourists have been killed by these incredible predators.
HOW Flores XP Adventure Tours (www.floresxp.com) offers a variety of tour options and prices.

BIG CATS
Tanzania

Tanzania's Serengeti is one of the greatest places to view lions, leopards and cheetahs, where they are relatively used to seeing tour groups and it's easy to get close. For something more visceral, try a walking tour in Katavi National Park, where you will literally be out there face-to-face. Your guide will be trained to carefully check the brush and surrounding landscape to make sure there are no cats lurking.
HOW Bukoba Tours (www.bukobaculturaltours.co.tz) is one of many reputable tour operators for the area.

GREY WOLVES
USA

In 1995, 31 Canadian grey wolves were brought to Yellowstone National Park in an attempt to reintroduce the species to its historic habitat. This resulted in what is likely the world's most famous wildlife reintroduction. While you're unlikely to get face-to-face with the notoriously elusive grey wolf, a spot from a distance is still a spine-tingling moment. Go in winter when there are fewer crowds and the snow makes spotting the animals easier.
HOW Head out on a tour with Yellowstone Wolf Tracker (www.wolftracker.com).

Left: A leopard surveying its domain

TRAVEL WITH

Actively trying to minimise your plastic use on the road forces you to travel more consciously, which will help to deepen your experience in your destination while limiting your impact on it.

Plastic-free packing list

Reusable bowl & mug
Sea To Summit (www.seatosummit.com) sells collapsible bowls and cups with lids (handy for street-food dining); the BPA-free plastic KeepCup (www.keepcup.com) is durable and barista-friendly.

Reusable cutlery
Kathmandu (www.kathmandu.com.au) makes a super-light titanium set of cutlery, while Hikenture (www.hikenturestore.com). sells an especially nifty foldaway set.

Water filtration bottle
Filtration bottles remove particles and bacterial nasties, as well as unpleasant tastes. Options include LifeStraw (www.lifestraw.com), and the Grayl bottle (www.thegrayl.com).

Water bottle
Opt for a plastic-free water bottle, or at least a BPA-free plastic one. Tree Tribe (www.treetribe.com) plants trees for each stainless-steel bottle sold, while sales of S'well bottles (www.swellbottle.com) aid Unicef water programmes.

Reusable carry bag
Opt for a bag made from sustainable fabric such as organic cotton. EcoRight (www.ecorightbags.com) sells a range of sustainable carry bags, while Australian company Boomerang Bags (www.boomerangbags.org) sells affordable bags made from recycled material.

Bamboo toothbrush
Limit landfill by opting for a more sustainable alternative to a plastic toothbrush, such as the Environmental Toothbrush

(www.environmental toothbrush.com.au) or Mother's Vault toothbrush (www.mothersvault.com), both of which are made from bamboo and BPA-free nylon.

Shampoo bar
Plastic-free shampoo bars provide far more washes than a standard travel-size bottle. Options include the all-natural Seaside Shampoo Bar by Clean Coast Collective (www.cleancoastcollective.org), and the #BeCrueltyFree bar by Lush Handmade Cosmetics (www.lush.com).

OUT PLASTIC

The devastating effects of plastic on the environment has inspired a new culture of plastic-free living, yet environmentally friendly habits can easily slip on holiday when the convenience of single-use plastics can be particularly tempting. But they don't have to. In fact, with a bit of preparation, it's now easier than ever to travel without plastic, and it's an incredibly rewarding feeling to depart a destination with the knowledge that you've done your best to minimise your plastic footprint on it.

There are some fantastic travel accessories out there to help you reduce your plastic footprint (though using what you already have should be your first option). Training yourself to actively avoid single-use plastics is also key. Get in the habit of asking flight attendants to fill your reusable water bottle rather than accepting drinking water from plastic cups, and politely decline plastic stirrers and straws in cafes and bars. By packing your own toiletries, you'll be less tempted to use the single-use options provided in hotel bathrooms, and if you invest in your own eating utensils (or simply take some from your drawer at home), you'll never get caught out having to use the disposable variety. To make it easier on yourself, opt for tour operators that provide potable drinking water with which to fill your own bottle, and support accommodations committed to limiting single-use plastics.

TAKE IT FURTHER
Do your bit:
Preserve the planet, p228
Get an education in climate change:
Learn about fragile places, p244

BALI, INDONESIA

To help combat the plastic crisis in Bali, which has poor waste management infrastructure, many tourism businesses now host beach clean-ups; check out waste-prevention collective One Island One Voice (www.oneislandonevoice.org) for upcoming events.
HOW On Gili Trawangan's north coast, Gili Eco Villas (www.giliecovillas.com) is mostly solar powered, has a waste-water treatment system, and is committed to recycling.

COSTA RICA

In 2017, Costa Rica announced its intention to become the first country to ban single-use plastics by 2021. By opting to bed down at a Cayuga Collection hotel, which offer filtered well water and bamboo straws, you can join the movement.
HOW Splurge on a stay at Lapa Rios, a luxury ecolodge (www.cayugacollection.com) that protects 1000 acres of Central America's last remaining tropical lowland forest, and also supports the local community.

RWANDA

Aside from having one of the strictest plastic-bag bans on the planet, Rwanda's cleanliness is largely thanks to Umuganda, a community clean-up held on the last Saturday of every month, when all able-bodied locals between the ages of 18 to 65 are required to pitch in.
HOW Adjacent to Volcanoes National Park, the home of Rwanda's mountain gorillas, Wilderness Safaris' Bisate Lodge (wilderness-safaris.com) is the nation's first eco-sensitive luxury safari camp.

AUSTRALIA

Although Australia has been slow to implement wide-scale single-use plastic bans, many grassroots organisations are working hard to make a difference. These international nonprofit Sea Shepherd Conservation Society (www.facebook.com/ssaubeachcleanups), which hosts regular beach clean-ups.
HOW Byron Bay YHA (www.yha.com.au) has a 'planet-friendly' ethos and makes use of rooftop rainwater tanks and solar panels.

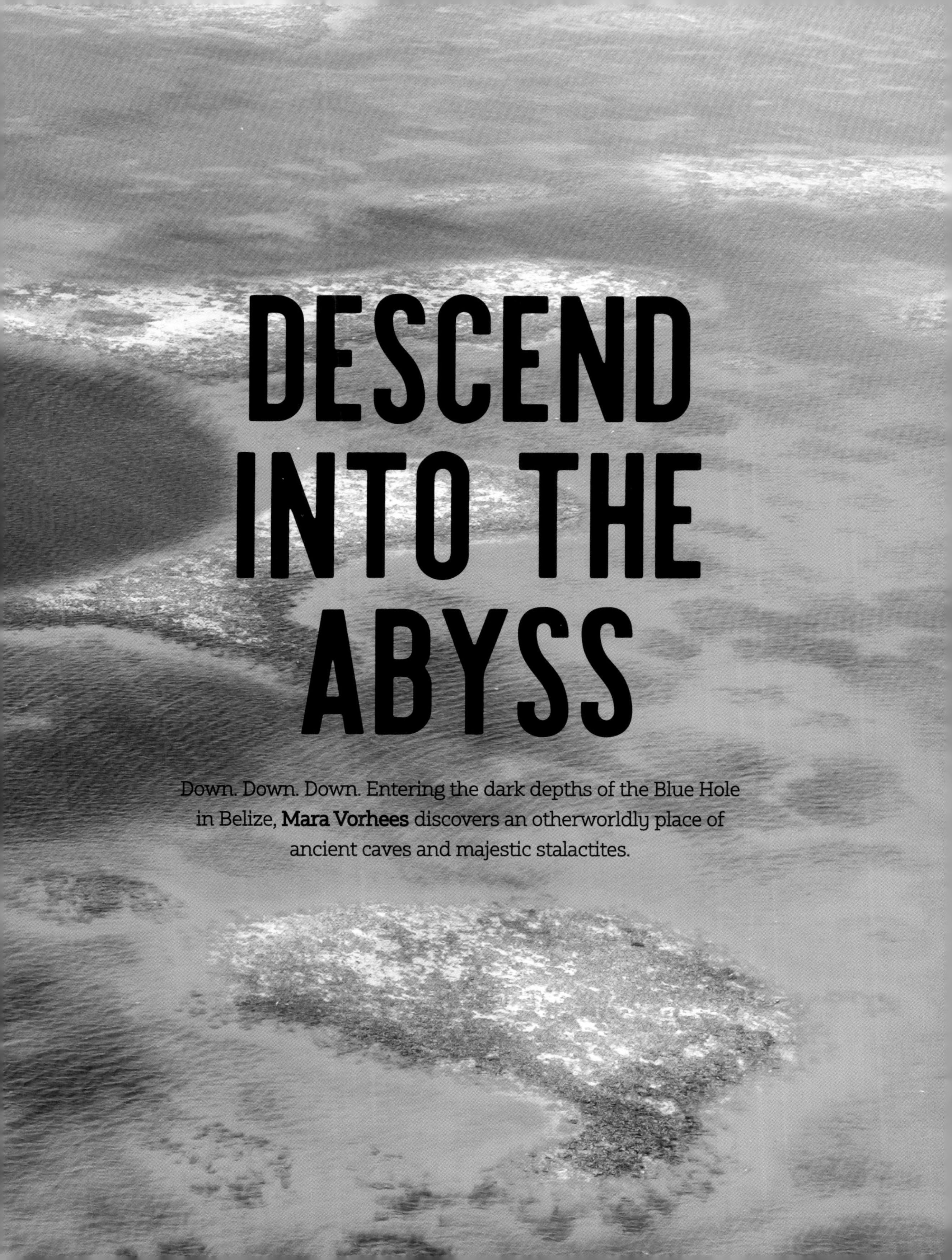

DESCEND INTO THE ABYSS

Down. Down. Down. Entering the dark depths of the Blue Hole in Belize, **Mara Vorhees** discovers an otherworldly place of ancient caves and majestic stalactites.

When Jacques Cousteau puts a dive site on his Top Ten list, then you know it must be good. That's what I was reminding myself as I descended along a wall into the depths of the Great Blue Hole, a Unesco World Heritage site off the coast of Belize.

So far, it was dark.

But, suddenly, the wall gave way to a cavern. I peered through the gloom to discern massive limestone columns suspended from the ceiling of the cave. This eerie, upside-down forest resembled nothing I had ever seen above or below the surface of the ocean. That's what motivates divers to cross the 43 miles (70km) from the Belize mainland to the Blue Hole: the chance to experience the otherworldliness of our amazing earth.

The Great Blue Hole is a giant sinkhole – a vertical cave – in the middle of the Lighthouse Reef. It's almost 1000ft (305m) across, so you can get a vague idea of its outline from the edge, though it's more easily visible from the air. And it's about 400ft (122m) deep, apparent from its rich, indigo hue, which contrasts so dramatically with the surrounding turquoise waters. From above, the Blue Hole is a union of light and darkness, clarity and obscurity, vibrancy and mystery. It beckons you to go deeper.

On the boat ride out, our dive guide recounted the history of this iconic site. It's basically a limestone cave that was formed during the last ice age. When the temperatures warmed and sea levels began to rise, the cave flooded and eventually collapsed, creating what is believed to be the largest submarine sinkhole in the world. The appeal, the guide reminded us, is not the sea life, which is sparse and difficult to see in the dark; rather, the cave system contains fascinating geological formations, unlike anything you'll see anywhere else.

No other boats were in sight when we motored through the cut in the reef and anchored at the southern mooring. Our group donned our gear and descended. Our first stop was a sandy ledge, which sloped down from about 18ft to 50ft (6m to 15m). Then we dropped off the ledge and into the blue abyss.

I concentrated on equalising, which always seems to require more time and effort from me than from my fellow divers, so I did not see much as we descended. But actually, there's not much to see except the limestone wall, stretching down, down, down, and out

of sight. The guide pointed out the reef sharks that swam in and out of our view. The light gradually faded, as at dusk. And we descended.

Before long, we reached 110ft (34m). It was dark, but I could make out the mouth of a sort of cave, studded by eerie, enormous stalactites. They were massive – 12ft (3.6m) in length, and much too wide for one person to reach around. We navigated among the majestic suspended structures, marvelling at their size, their age, and how they came to be in the mysterious depths of the ocean. Like the giant redwoods of California or the great cathedrals of Europe, the Great Blue Hole was an almost holy experience (unless that was the nitrogen narcosis at work).

Then the guide tapped on my tank, indicating that our time was up. Maximum depths of 130ft to 140ft (40m to 43m) mean that bottom time is limited to about eight minutes. It was too short, for sure; but it left me filled with wonder at the enormity and geodiversity of our planet. Here I was in one of her deepest, darkest and most foreboding corners – surely a place not meant to be seen by human eyes – but even here she showed off an unexpected majesty.

HOW The Great Blue Hole requires Advanced Open Water Diver certification; most operators also ask that divers to do a local 'check-out' dive first. Based on nearby Long Caye, Huracan Diving (huracandiving.com) organises dive packages that include the Great Blue Hole, while dive shops in Caye Caulker and San Pedro also offer day trips.

"The guide pointed out the reef sharks that swam in and out of our view. The light gradually faded, as at dusk. And we descended."

Above and left: Waitomo Caves, New Zealand; divers exploring the Blue Hole.
Previous page: The Blue Hole is a huge underwater sinkhole off the coast of Belize

WIELICZKA SALT MINE
Poland

Up to 1000ft (305m) below the earth's surface, this Unesco World Heritage site was a source of precious salt for seven centuries. Today, the vast underground network contains a museum, four spectacular chapels and countless sculptures, all hewn out of salt. It's an incredible example of artistry and productivity in the unlikeliest of places.
HOW Wieliczka is 9 miles (15km) south of Kraków, accessible by a 20-minute ride on the regional Koleje Małopolskie train. See the highlights on a 2-mile (3.5km) walking route, or soak up the salty goodness at the onsite health resort (www.wieliczka-saltmine.com).

LOST WORLD CAVE
New Zealand

Enter through a narrow crack in the earth's surface and descend 330ft (100m) into a majestic underground cavern, abseiling into the abyss. The Lost World is unlike other cave systems – it is filled with greenery and growth evoking the prehistoric world. At the bottom, explore caverns, gaze at glow-worms and admire the planet's incredible architecture at its extreme and remote best.
HOW Waitomo Caves are located in King Country on New Zealand's North Island (about 90 miles/150km west of Rotorua). Abseiling tours can be booked through Waitomo Adventures (www.waitomo.co.nz).

THRIHNUKAGIGUR VOLCANO
Iceland

Only in Iceland is it possible to descend into a dormant volcano, riding an open-air elevator 394ft (120m) down into the mountain's magma chamber. Thrihnukagigur is a natural anomaly, as a volcano's magma normally cools and solidifies, sealing off the crater when the volcano goes dormant. In this case, however, the magma drained away, leaving behind a vast open chamber, the walls an incredible canvas of colours and patterns.
HOW Inside the Volcano (www.insidethevolcano.com) operates Thrihnukagigur volcano tours from May to October; pickup from Reykjavík is included.

Sometimes the mere mention of a country conjures up images of war, famine or other unbelievable suffering. But a place and its people can evolve in a generation. Venturing into misunderstood lands can be an enlightening experience, teaching you as much about the spirit of a destination as yourself and your own preconceptions.

CHALLENGE YOUR PERCEPTIONS

BUILD UP TO IT
Be open to friendship:
Accept the kindness of strangers, p44
Discover new stories:
Live on the fringe, p190

Left: *Looking out over Guatape, Colombia*

Childhood memories of a foreign land in strife can stick with us long after any pity is warranted. Wars end, crops recover, and natural disasters do finally become a tragic chapter in a history book. Visiting places with troubled recent histories will often have friends and family shaking their heads and issuing stark warnings, but – and while it is wise to stay informed – these kinds of trips can also broaden our horizons in unexpected ways.

Visiting a country synonymous with war and discovering jubilant nightlife and welcoming people will uplift and enlighten in equal measures. Venturing to a place expecting poverty and hardship and instead finding a thriving art scene and fantastic food can't help but educate and change our view of developing countries. The surprising often also makes us feel more alive – it stimulates the nervous system, giving us a jolt of novelty and intrigue, all the while imparting lasting memories. And even more than this, surprising situations force us to be curious and look at things in new ways, including our own prejudice, and as our viewpoint morphs when faced with reality, we become inspired to share our findings with friends, colleagues and families. Challenging your perceptions is not only exhilarating and educational, it also makes the world a richer place.

RWANDA

The population of gorillas in Rwanda has tripled in the last 15 years. The (hefty) national park fee helps insure their safety from poachers and contributes to a growing economy in a land once stained with genocide. Hike through the cloud forest, clinging to the slopes of a volcano, to glimpse gorillas in their natural habitat. **HOW** Try tour operator Hills in the Mist Tours (www. hillsinthemisttours.com). The dry season runs June to September, December and January, but since this is a cloud forest, it's probably going to rain anyway.

SIERRA LEONE

A country ravaged by war and visited upon by a biblical Ebola outbreak is now the darling on the West African travel circuit thanks to palm-dappled beaches (River No 2 beach is popular), abundant wildlife, a revived capital and an aromatic kitchen. **HOW The Place beach resort (www.stayattheplace.com) is set on beautiful Tokeh Beach, a white-sand jewel backed by Western Area National Park, near the point of the Freetown Peninsula. Fly into Freetown via Paris, Brussels or Nairobi.**

TUNISIA

A 2015 ISIS attack resulted in 30 dead British tourists. Security has ratcheted up, yet Tunisian attractions – divine beaches, vast Saharan dunes, and an old Star Wars set (Tatooine) – remain empty. And the city of Tunis, with its bohemian neighbourhood, Sidi Bou Said, set along the Mediterranean is a stunner. **HOW Fly to Tunis-Carthage International Airport from Cairo, Dubai, Amman, Frankfurt, Barcelona and Paris, among others.**

COLOMBIA

Colombia is no longer a cartel stronghold. Capital city Bogotá is thriving, along with Cali, Cartagena and Medellín (once the murder capital of the world), and with the recent FARC peace agreement, the Colombian Amazon is ripe for visitors too. Explore Andean summits, unspoiled Caribbean coast, enigmatic Amazon jungle, and cobbled colonial communities. **HOW Fly to Colombia from New York, Madrid or Mexico City.**

There are a few spots on earth so majestic, so spellbinding, that to clap eyes on them in real life is to change forever. They may be as familiar as your own face from a thousand pictures, but they can't disappoint, they just weren't made that way.

BEHOLD
A TRAVEL
ICON

ULURU
Australia
Nothing prepares you for the immensity, grandeur, changing colour and stillness of 'the Rock'. Uluru is 2.2 miles (3.6km) long and rises 1142ft (348m) from the surrounding scrubland. If that's not impressive enough, it's believed that two-thirds of the rock lies beneath the sand. **HOW** Uluru-Kata Tjuta National Park is 278 miles (447km) from Alice Springs.

BOROBUDUR TEMPLE
Indonesia
Dating from the 8th and 9th centuries, Borobudur is the world's largest Buddhist temple and one of Indonesia's most important cultural sites. **HOW** Borobudur lies 26 miles (42km) northwest of Yogyakarta; it is open from dawn to dusk daily.

GRAND CANYON
USA
The canyon's immensity, the intensity of light and shadow at sunrise or sunset, even its very age, scream for superlatives. At about two billion years old – half of earth's total life span – the layer of Vishnu schist at the bottom is some of the oldest exposed rock on the planet. **HOW** The most accessible area of the park is the South Rim, a 60-mile (96km) drive north of I-40 at Williams, Arizona.

TAJ MAHAL
India
The Taj Mahal rises from Agra's haze as though from a dream. Seeing it in person, you'll understand it's not just a famous monument, but a love poem composed of stone. When you first glimpse it, you might find yourself breathless with awe. **HOW** The Taj Mahal is in Agra, a two-hour train journey from Delhi; it is open dawn to dusk, Saturday to Thursday.

BUILD UP TO IT
Enter a spiritual realm:
Seek out sacred places, p46
Be at the frontier of human ambition:
See a rocket launch, p176

© DAVID ASMUTH | GETTY IMAGES

Left: Grand Canyon National Park, Arizona, USA

Boost the already considerable benefits of going for a long-distance walk by choosing a trail that crosses an entire nation – get to know every nook and cranny and feel the thrill of knowing you've left no mile untrodden.

Walking makes you feel good: the fresh air and physicality, the slow speed, the momentary escape from the clamour and chaos of daily life. There's no doubt about it, going for a walk is wonderful for mental well-being. So imagine the boost your spirits can get from walking across an entire country.

Naturally, countries vary in size, which means there are options for walkers of all abilities, from Luxembourg-lite to the hardcore expanse of Canada. But the thrills of following a route are many. There is no better way to get to know a nation than by hiking right across it. Meeting the people of its uplands, lowlands, cities and steppes; watching its accents,

menus, topography and traditions change – maybe subtly, maybe dramatically – as it stretches out between its own borders.

Then there's the sense of achievement. By their nature, cross-country trails tend to take at least a few weeks; they could take months or years. That means you get to feel the matchless freedom and the growing fitness that only comes from a multiday, self-propelled journey. And it means that when you reach the end – when you dip your tired toes in the frontier-defining ocean or arrive at the next customs post – you can bask in the satisfaction of true completion. Though this powerful sensation may make you want to just keep going across the next nation...

TRAVERSE A COU

COAST TO COAST
England

Not the quickest way to cross the UK – following Hadrian's Wall is shorter – but the Coast to Coast trail, linking the Irish and North Seas, is a classic. Brainchild of hill-walking hero Alfred Wainwright, the C2C runs between St Bees on the west coast and Robin Hood's Bay on the east, traversing three national parks along the way.

HOW Accommodation en route can get booked up well in advance. Book ahead, especially in smaller places. See www.wainwright.org.uk.
Distance: 192 miles (309km).
Duration: 11–14 days.

JORDAN TRAIL
Jordan

Running the length of Jordan, this trail traces the Great Rift Valley, winding via Muslim and Crusader castles, wild wadis, nature reserves, Bedouin-roamed desert and archaeological wonders. If short on time, try the 45-mile (73km) section south from Dana, to finish in Petra.

HOW Accommodation varies on route. Be ready for nights in the wilderness, or book an organised trip. See www.jordantrail.org.
Distance: 400 miles (650km).
Duration: 35–40 days.

TE ARAROA
New Zealand

Terrific Te Araroa (www.teararoa.org.nz) spans the North and South Islands, encompassing coast, capital, Southern Alps, fizzing volcanoes and everything in between. Start hiking south from Cape Reinga in October or November to try to get past the biggest mountains before the snow hits.

HOW As well as huts, campsites and B&Bs, Trail Angels offer a bed for the night – see trailangel.co.nz.
Distance: 1850 miles (3000km).
Duration: 100–160 days.

Below: Walking through the Lake District National Park on England's Coast to Coast trail

NTRY ON FOOT

© MONKEY BUSINESS IMAGES | SHUTTERSTOCK

BUILD UP TO IT
Walk with purpose:
Make a pilgrimage, p60
Reach for higher ground:
Summit a mountain, p152

MEDITATE WITH MASTERS

You know it because you feel it. And when you come into contact with someone who exudes equanimity, whose wisdom and love overflow, there's a good chance they are regular practitioners of meditation.

Yoga studios can be found everywhere you go, the benefits of a regular meditation practice are well known, and there have been enough mindfulness studies to fill entire volumes of medical journals — libraries even. We know we should do our yoga, meditate daily, and that retreats are the best way to kick-start a deeper practice. But with so many teachers and retreats available, you'd do well to seek out the masters of the form. The elders, the experienced mystics who deliver a deeper wisdom. That's not to say the bendy new model yogi doesn't have its virtues, but there's nothing like mainlining the meaning of life from someone who has been studying the sutras since the Beatles were still a band.

GURU SINGH

Los Angeles, USA

Known to blend Eastern mysticism with Western pragmatism, Guru Singh was one of Yogi Bhajan's (see panel, *right*) first students. He offers retreats around the world as well as weekly classes and regular workshops at Yoga West in Los Angeles. He teaches what he calls conscious living, urging students towards a more enriched and enlightened life through the practice of kundalini yoga, meditation and sacred sound. He teaches year-round.

HOW: A two-night retreat with Guru Singh (www.gurusingh.com) at Kripalu Center for Yoga & Health, includes tuition, accommodation and meals.

THICH NHAT HANH

Plum Village, France

The mindfulness meditation master and legendary peace activist Thich Nhat Hanh is based at Plum Village (www.plumvillage.org), 53 miles (85km) east of Bordeaux, France, where he teaches, writes and speaks. Guests are welcome as day visitors or to attend 90-day retreats, which is the best way to come across the Zen Master himself. Autumn (September to December) is ideal.

HOW Plum Village is best reached via the Bordeaux-Mérignac airport. Programmes include lodging, food and teachings.

VIPASSANA MEDITATION

Myanmar (Burma)

Vipassana meditation courses are taught at retreat centres all over the world and cost nothing at all. You must apply to attend a 10-day introductory course, during which you cannot speak, write or read a single word. Courses are run solely on a donation basis, contributed by those who have taken the course previously and want to give back. Teachers are not paid. And yet the movement grows.

HOW Myanmar Vipassana Meditation Centre (www.joti.dhamma.org) is a good place to learn. Visit the website to read the Code of Discipline and complete an application form to get started.

Left: Thich Nhat Hanh at Plum Village, France

Kundalini Rising

In the 1960s, Yogi Bhajan, a relatively young Sikh from Amritsar, India, arrived in Los Angeles driven to bring Kundalini yoga, a little-known tradition practised by Sikh yogis for centuries, into the mainstream. He opened an antique-furniture store in Los Angeles where he taught hippies who piled in to learn at his feet. What he offered was a mixture of mantra (chanting), asana (posture) and meditation that he promised would redirect their attention and energy to make them more aware and effective in their lives. He called it the yoga of householders. Today, it's taught all over the world.

BUILD UP TO IT

Feel blissfully buoyed:
Spend time on water, p14
Dedicate time to your self-care:
Retreat, p106

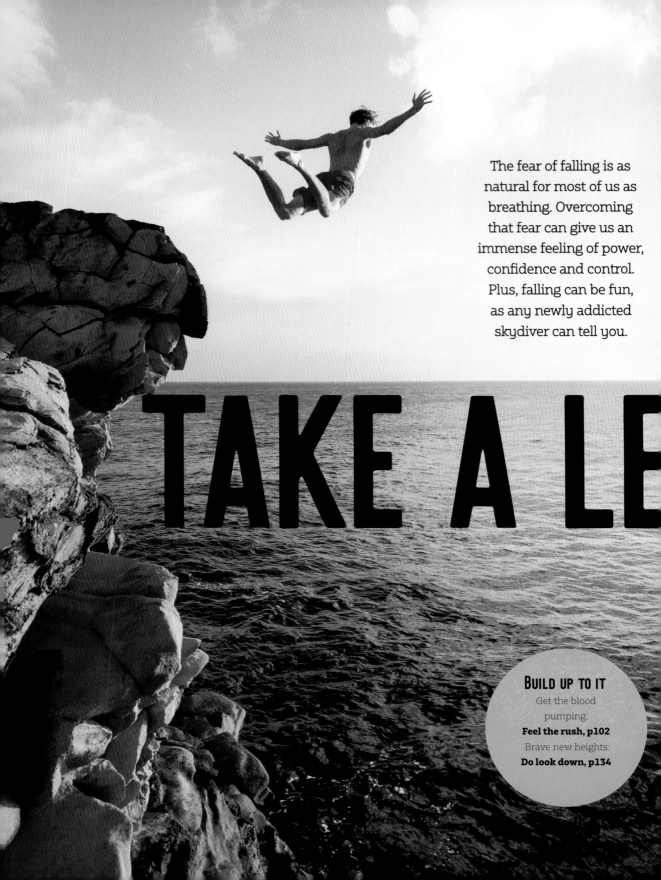

The fear of falling is as natural for most of us as breathing. Overcoming that fear can give us an immense feeling of power, confidence and control. Plus, falling can be fun, as any newly addicted skydiver can tell you.

TAKE A LE

BUILD UP TO IT

Get the blood pumping:
Feel the rush, p102
Brave new heights:
Do look down, p134

Left: Cliff jumping at sunset

If you're the kind of person for whom bungee jumping is just another Saturday afternoon, stop reading. But if you're like most of us – wary of heights, shying away from cliff edges and anything involving a parachute – please listen. We're about to suggest you do something scary: jump out of a plane, zip through the sky on a harness, tiptoe over a chasm.

Why would I want to do that, you say? Well, research suggests that adventure sports, many of which involve overcoming the fear of falling, have both physical and emotional benefits. A recent study led by British psychologists concludes that 'even the most extreme adventurous physical activities are linked to enhanced psychological health and well-being outcomes'. Despite the traditional perception of risk-takers as deviant thrill-seeking bros, the study shows that extreme sports enthusiasts are men and women from all walks of life. What they share is a desire to change themselves. And, according to the research, they do. Benefits of extreme adventure activities include enhanced quality of life, better emotional regulation, greater emotional agency in interpersonal relationships, feelings of joy, higher levels of social connections and a sense of overcoming fear. Um, yes please.

It's OK to start small. Try a less-extreme activity first, and work your way up. Ziplining, say, before skydiving. Or if you want to make a (literal) giant leap into the abyss, strap on your parachute and go for it.

AP OF FAITH

BEGINNER: STOMACH-DROPPING SWING

Amsterdam, The Netherlands
Atop north Amsterdam's A'Dam Tower you'll find Swing Over the Edge, a huge, vertigo-inducing swing suspended over the side of the building's roof deck. Swing out into the chilly Dutch air, gawping down at the blue-grey harbour and the gabled roofs of the canal houses. From 330ft (100m) high, you'll feel like an all-powerful giant.
HOW A'Dam Lookout (www.adamlookout.com) is open 10am to 10pm daily.

INTERMEDIATE: 007 BUNGEE JUMP

Switzerland
In the 1995 movie *GoldenEye*, Pierce Brosnan's James Bond leapt from the edge of the Contra Dam over Switzerland's Verzasca Valley. The spot is now home to the popular '007 jump', a 722ft (220m) bungee experience for the brave at heart. Shuffle to the lip of the vast concrete dam, taking in the mountains of Italian-speaking Ticino, then... jump.
HOW Jump any time from Easter to October. See www.trekking.ch/en.

ADVANCED: DEEP WATER SOLOING

Mallorca, Spain
In Mallorca, the homegrown sport of deep water soloing – rock climbing without a rope, then leaping into the ocean below – is called 'psicobloc', which means 'psycho bouldering'. Indeed, there's something crazy (in a good way!) about ascending a sun-beaten limestone cliff in nothing but a swimsuit, then plunging into the warm Mediterranean below.
HOW Rock & Water Mallorca (www.rockandwatermallorca.com) offers guided climbs.

EXPERT: SKYDIVING

Queenstown, South Island, New Zealand
If you're gonna skydive, do it here, in the so-called adventure capital of the world. Plummet from up to 15,000ft (4500m), trying your best to take in the snowy peaks and jade lakes of the Wakatipu basin as you fall. Bring your adrenaline down afterwards with a beer at one of the city's famous pubs.
HOW NZONE Skydive (www.nzoneskydive.co.nz) offers tandem skydives.

Is who we are limited to where we come from? Do we retain the same identity for life? **Omo Osagiede**'s travels through Morocco provided an opportunity to reflect on the question of how travel can change the very notion of who you think you are.

CHANGE YOUR IDENTITY

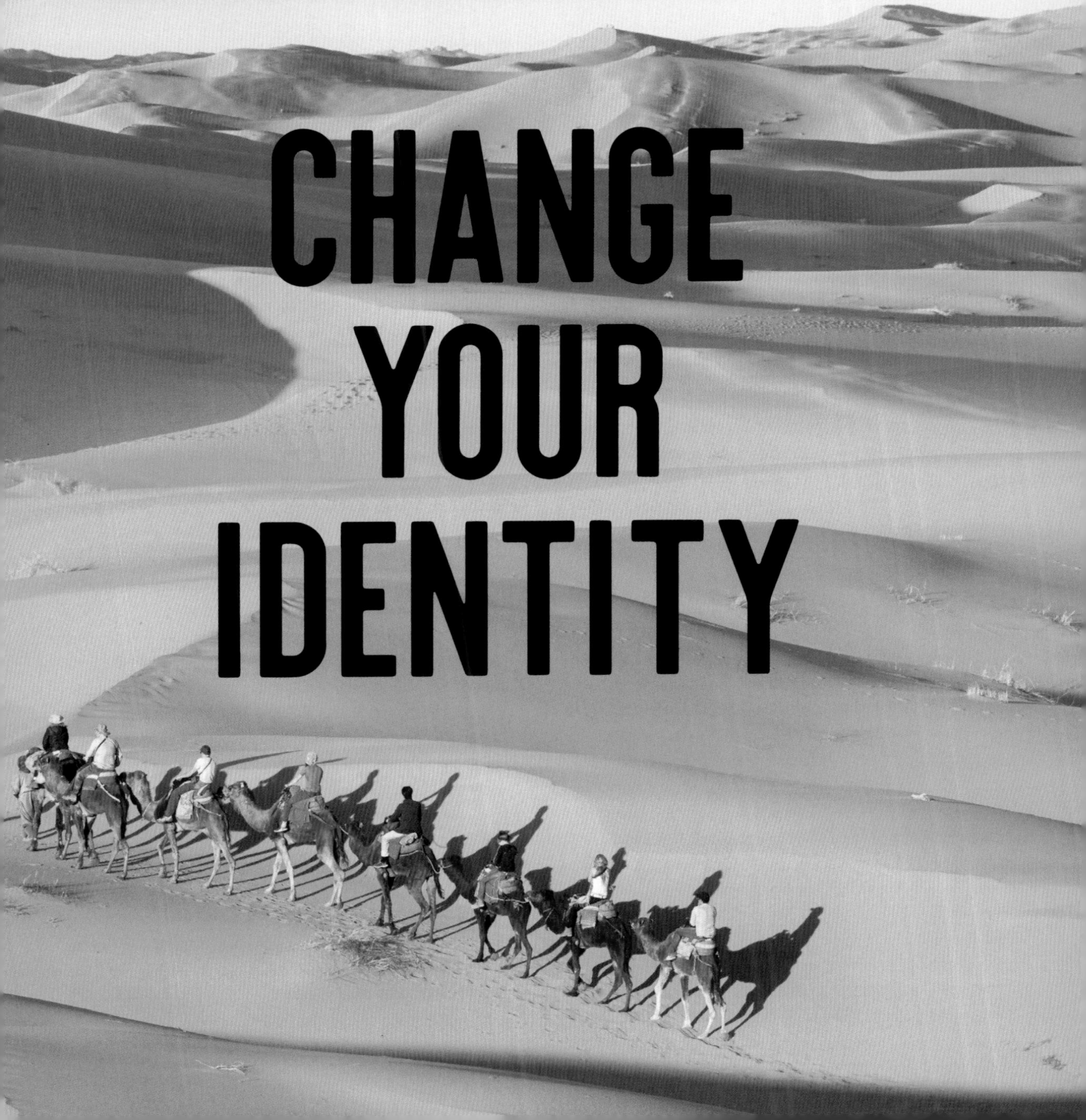

أخي أخي, لا تخبئ عنّا الأشياء الجيدة, شاركها معنا من فضلك.

("*Brother! Brother! Don't hide the good stuff from us! Share it with us please!*"), a young man whispered, looking towards me. I could barely make out his face in the soft glow of the lamps in the dinner tent. Oblivious to the fact that I was the one being addressed, I carried on my conversation.

Less than two hours before, I had experienced the most amazing sunset I had ever seen. Witnessing the sun cast its golden glow over the Sahara Desert as it hung low in the horizon, I suddenly understood why our driver had been in a mad rush to get through our itinerary for the day. At that moment I only had one sentiment: one of gratitude to be alive on New Year's Eve to witness the last sunset of the year surrounded by red desert sand that had swallowed kingdoms and given birth to others.

يا أخي, أخي, نحن نتحدّث إليك!

("*Brother! Brother! I'm talking to you!*"), the young man entreated. He had been joined by another. Their earnest faces had taken on a look of confused curiosity. They could not understand why I, who was dressed from head to toe like a Berber, was ignoring them.

أرنا أين يوجد البرتقال لو سمحت يا أخي.

("*Show us where the oranges are please brother*"), they begged. By this time they had gained my full attention. I returned their look of incredulity. Suddenly the truth hit all of us at the same time.

"Ah! Brother! You don't speak Arabic? We thought you were Berber!" They burst into fits of laughter.

"Brother! We're so sorry! Where are you from?"

"Nigeria," I replied, humoured and secretly proud that I blended in well enough to pull off the second case of mistaken identity in one night. Previously, a British lady had also mistaken me for a Berber guide asking if I could help her daughter tie her *shesh* (traditional turban worn by the nomads). The young men hugged me and amidst a flurry of high fives and back slapping, I learned that they had travelled from Casablanca to spend New Year's Eve in the Sahara Desert. There was clearly something special about the place that attracted even the locals.

I grew up in Africa, at a time when the continent was still struggling to cast off its cloak of confusion caused by colonialism, poor leadership and endless conflicts. That backdrop combined with my Nigerian heritage helped shape my understanding of my African identity. From a political perspective, the Africa of my childhood was cast in the shadow of legendary figures including Obafemi Awolowo, Thomas Sankara, Kwame Nkrumah, Julius Nyerere, Kenneth Kaunda, Archbishop Desmond Tutu and the great Nelson Mandela. History books brought these characters to life, creating for me an impression of a predominantly black Africa that was once liberated but now 'burdened' by independence.

From a literary perspective, gifted African authors including Chinua Achebe, Ngũgĩ wa Thiong'o, Cyprian Ekwensi and Wole Soyinka further shaped my impressions about Africa, its peoples, its cultures, its pain and its beauty. However reading their stories, seeing the images projected by mass media and being raised in Nigeria did not necessarily mean I fully understood what being African meant. I was African but my knowledge of Africa was incomplete. The missing ingredient to my understanding of 'African-ness' was travel.

In Morocco, my African identity was reborn. Travelling through the mountains, valleys and deserts there helped me appreciate the rich diversity, beauty, colour, character, geography and history of my continent of birth in a way I never had before. As we drove past village after village of red-mud-brick buildings built around free-standing minarets hidden in the mountains, I began to weave mental threads between Berber culture and the nomads of Northern Nigeria.

Although 'my Africa' was born on the shores of the Atlantic Ocean and rooted in the forests of the ancient Benin Empire, it was now evolving as we travelled through Sahara Desert dunes and the Low and High Atlas, to discover itself on the Moroccan shores of the Mediterranean. I discovered that my Africa is black but it is also white, brown, Berber, Amazigh and Arab.

In Morocco I saw the imperfections of my African identity reinforced. However, I also saw its progressive nature. I saw a people who mirrored the unbowed 'thrive-in-the-midst-of-difficulty' spirit that is represented in every country on the African continent.

A Ghanaian blogger once said: 'Identity is fickle. Identity isn't set in stone, it is ever changing, and no one is ever enough of anything...'

... And I am OK with that.

> *"I was African but my knowledge of Africa was incomplete. The missing ingredient to my understanding of 'African-ness' was travel"*

Above: Village in the High Atlas Mountains, Morocco.
Previous page: Traversing the Sahara Desert

TREK WITH BERBERS
Morocco

Morocco's proud indigenous people are a memorable part of many travellers' journeys in the country. Their Amazigh colour and character are a big part of special spots such as Marrakesh, the Atlas Mountains and the Sahara. Try the nomadic lifestyle in the desert sandscapes of Erg Chebbi and the Draa Valley. **HOW** Berber-British partnership Wild Morocco (www.wildmorocco.com) runs three- to six-day treks into the Atlas Mountains and Moroccan Sahara following nomadic migration routes.

CROSS CONTINENTS
Turkey

Fate has put Turkey at the junction of two continents. A land bridge, meeting point and battleground, it has seen many people – mystics, merchants, nomads and conquerors – moving between Europe and Asia since time immemorial. In İstanbul, you can board a commuter ferry and flit between Europe and Asia in under an hour. **HOW** On the European side, the major ferry docks are at Eminönü and Karaköy. Buy a ticket and board a ferry for the bustling district of Kadıköy on the Asian side.

BE THE TOUR GUIDE
Worldwide

Showing visitors around your home town has the uncanny knack of keeping the wanderlust alive when you're not travelling; but it also makes the place where you live an intricate part of your make-up. There's something about sharing your passion with strangers – be it a love of your local artisans or urban wildlife – that crystallises it inside. **HOW** Tours By Locals (www. toursbylcoals.com) connects local guides with travellers. Curate your own personal tour and join up to spread the love.

SURVIVE IN TH

Survival in the wilderness

Andrew Thomas-Price, wilderness instructor at
Dryad Bushcraft (www.dryadbushcraft.co.uk), shares
his top tips for survival in the wilderness.

1. Stay healthy and hydrated

Rock boiling is a method of purifying water, removing viruses and bacteria. Heat the rocks in a fire and drop them into the water. The residual warmth from the rocks will heat the water rapidly and eventually bring it to the boil.

2. Protect your core temperature

This differs depending on where you are – in the desert you might dig a shallow depression to create shade, in woods you might build a debris shelter, using sticks to make the frame and dead leaves for insulation (or an open-fronted lean-to if you have a fire). In cold climates, you might dig a snow trench or cave.

3. Know where you are going

It's possible to navigate using the sun and stars. At night, you might look for constellations like the North Star and Southern Cross. During the day, you could put a stake in the ground, mark where the shadow lies, and then wait 20 minutes and do it again: the point directly between the two shadows is south (behind you is north).

4. Borrow from nature's larder

When foraging for nettles, sea herbs, vegetables, shellfish, berries and mushrooms, only take what you need. As a rule, the weirder a mushroom looks, the more edible it is – but never take chances. Medicinal herbs are also found in the wild, from meadowsweet (a natural aspirin and antiseptic) to plantain (for cuts and sores) and water mint for aiding digestion.

WILDERNESS

It helps to have some survival techniques under your belt when striking out into the wild on your travels – the most basic skills, from fire lighting to shelter building, could just save your life.

Survival: it's the most basic of human instincts. But in a world where everything happens at the flick of a switch, we've largely lost touch with the skills that once helped us in fight-or-flight mode. While we've all seen survival on the screen – in films such as *Into the Wild* and *The Revenant*, or Bear Grylls' and Ray Mears' TV shows – the reality can hit when we least expect it, especially when we travel out of our comfort zone.

Being prepared for the worst-case scenario is key when heading into wild places: whether you're climbing mountains, crossing a desert, navigating in the forest or traversing rivers, snowfields and glaciers. Then the simple rules-of-thumb and emergency kit come into their own, from being able to read a map to knowing how to use a rope or snow shovel, build a shelter, light a fire (without matches!) and, perhaps most important of all, keeping calm and devising a plan.

DRYAD BUSHCRAFT

Wales

Dryad Bushcraft leads survival courses with the coastline and ancient woodlands of the Gower Peninsula in South Wales as a backdrop. Skills learned on the basic course include fire lighting, identifying medicinal plants and constructing an emergency shelter. More advanced courses include foraging and trapping. **HOW Introductory and multiday survival courses range from beginners to advanced level. See www. dryadbushcraft.co.uk.**

AUSTRALIAN SURVIVAL INSTRUCTORS

Australia

On the New South Wales coast, an hour's drive north of Sydney, the Australian Survival Instructors' courses are terrific primers for a foray into the hot and harsh Australian outback. **HOW A four-day comprehensive course covers everything from aboriginal shelter construction to bush remedies and food, snares and traps and signalling methods. See www. aussiesurvivalinstructors. com.**

MOUNTAIN SHEPHERD ADVENTURE SCHOOL

USA

Based in Catawba, Virginia, Mountain Shepherd offers Survival 101 courses run by US Air Force SERE instructors, who teach you the seven priorities of survival, from injuries to finding and purifying water and improvising shelters. Other courses range from backpacking essentials to wilderness first aid. **HOW The overnight Survival 101 course includes accommodation in a lodge, tent or improvised shelter. See www.mountainshepherd.com.**

BUILD UP TO IT

Get in tune with nature:
Sleep under the stars, p42

Build your resilience:
Have an epic travel fail, p140

Left (above and below): Three Cliffs Bay on the Gower Peninsula, Wales; building a campfire

WATCH A TOTA

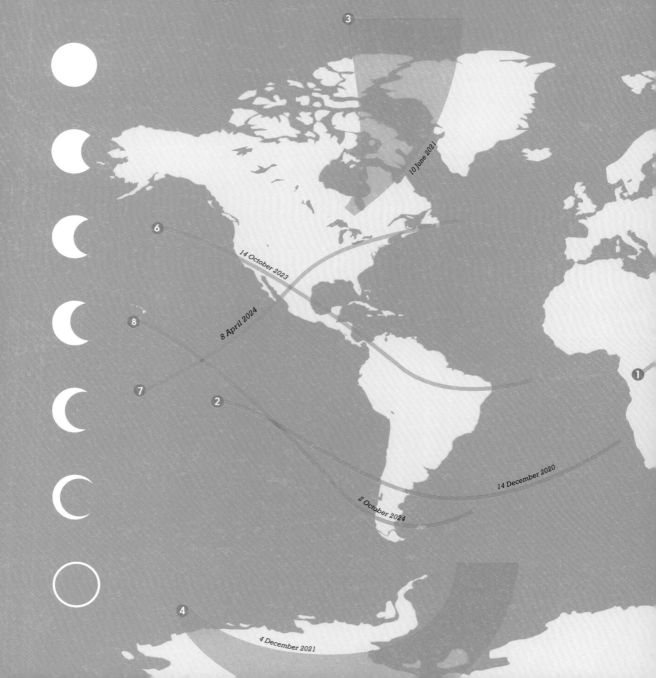

3

6 — 14 October 2023

10 June 2021

8 — 8 April 2024

7

2

4

14 December 2020

2 October 2024

4 December 2021

SOLAR ECLIPSE

10 June 2021

21 June 2020

20 April 2023

Nothing reminds us of the celestial ballet unfolding around us as watching a total solar eclipse. And with a little planning, it doesn't have to be a once-in-a-lifetime event.

On 21 August 2017, the Great American Eclipse became what's understood to be the most-watched eclipse in history. The path of totality cut a route across 14 states from Oregon to South Carolina and around 88% of the country's population gathered in fields and front yards (or online) to watch as the sun, moon and earth became aligned like giant snooker balls to cast their eerie shadow over those below. The experience? Pure wonder.

While the 2017 eclipse had no match in the USA for almost 100 years, if you missed it, there is no need to feel bound by the dance of the universe: find your own. A total solar eclipse will happen somewhere on earth every year or two. On 14 December 2020, a path of totality will cross southern Chile and Argentina, and on 4 December 2021, parts of Antarctica – an otherworldly destination in itself – will experience the same.

Travelling for a solar eclipse may seem extreme, but when combined with a longer trip, it allows us all to experience this awesome astral phenomenon at least once. And when the birds go silent, the air cools, the daylight dims and the crowds cheer, you'll feel the indescribable magic of being at once united with your fellow humans and acutely aware of your place in the universe.

PRESERVE THE PLANET

HELP A NATIONAL PARK
USA

The numerous national parks of the USA contain some of the world's most awe-inspiring natural wonders: the geysers of Yellowstone, Yosemite's waterfalls, the Grand Canyon... You can get involved with the National Park Service's conservation, research and restoration efforts by volunteering, donating or becoming a 'friend' to one of the parks. **HOW** Find out how to support the parks at www.nps.gov.

SAIL WITH PURPOSE
Seas and oceans

Document orca and humpback whales, monitor plastic debris, study remote ocean ecosystems – you can join an ocean research expedition and take part in marine exploration and conservation. As a crew member, you'll also learn sail rigging, navigation, weather analysis and all the basic duties of running an ocean vessel. **HOW** Sail away on the *Sea Dragon*, a 72ft yacht run by Pangaea Exploration (www. earth-changers.com).

SAVE THE LEMUR
Madagascar

The lemur, one of the world's most-endangered mammals, is facing 90% extinction within 20 years. Diverse specimens are endemic to Madagascar, an important biodiversity hotspot but also a very impoverished nation. Local conservation efforts could use some help. **HOW** Visit the Lemur Conservation Foundation or learn more about volunteering at www. lemurreserve.org.

TAKE PUBLIC TRANSPORT
Worldwide

Greenhouse gas emissions caused by airplanes are a major contributor to climate change; avoiding plane travel is the sustainable choice. Instead, take the bus or train and savour slow travel, watch the landscape gradually unfold and pop in on the locals along the way. **HOW** Spain's Feve narrow-gauge rail network starts in Bilbao and dawdles on through to Galicia. Book online at www. renfe.com.

DRIVE THE ELECTRIC HIGHWAY
Sweden

Visit the electric future, where a human-scale slot-car track is the first step towards a new reality free from the grip of fossil fuels. The 1.25-mile (2km) electrified stretch of road, between Stockholm Arlanda Airport and the Rosersberg logistics site, is a test run for a project that would see the technology rolled out across the country's highways. **HOW** Find out more about the project at www.eroadarlanda. com.

BUILD UP TO IT
Take the pressure off:
**Embrace the
off season, p12**
Support those who care:
**Join a conservation
project, p142**

At home, we do our bit to turn back the clock on the destruction of the earth – why stop when we travel? Whether you plunge into a hands-on conservation effort or just travel more mindfully, it's all part of the most important mission of our lifetimes.

SAFARI RESPONSIBLY
Kenya
Going on safari can be a thorny choice. The privilege of experiencing stunning wildernesses means visiting some of the world's least developed countries. Choose conscientiously – responsible operators work closely with local communities and conservation efforts to ensure the benefits of your travel dollar go where most needed.
HOW Ensure operators commit to community, habitat and wildlife, as does Gamewatchers Safaris (www.porini.com).

SLEEP IN A TREEHOUSE
Nam Ka National Park, Laos
You get a different perspective on the forest when you're living in the treetops. In Nam Ka National Park you can overnight in tree huts 130ft (40m) above ground, and soar above the jungle canopy on 9 miles (15km) of ziplines. Reforestation and sustainable agriculture initiatives here are run in partnership with local communities.
HOW The Gibbon Experience (www. gibbonexperience.org) offers a three-day stay.

TAKE A VEGAN TOUR
Europe
Lest you think it impossible to be both foodie and vegan, tours of cuisine-forward cities prove that taste-bud-tempting travel can be animal-product-free. As animal agriculture is responsible for 18% of greenhouse gas emissions (more than double that of aviation), a meat-free trip could end up being your ticket to vegan conversion.
HOW Head to Lisbon, Barcelona, Amsterdam, Rome and London with the help of Vegan Food Tours (www. veganfoodtours.com).

DIVE WITH A GOAL
Belize
With its beautiful coral reefs, teeming reef fish and famous Blue Hole, Belize is one of the world's favourite dive spots. Explore the dazzling deeps, 40% of which are protected, while helping to conserve them. Combat invasive lionfish, save sharks, care for injured manatees – there are loads of ways to help Belize's watery wildlife.
HOW See Earthwatch Institute (www. earthwatch.org) or Wildtracks (www. wildtracksusa.org).

TREK WITH HEART
Nepal
The home of Everest and Annapurna bears the burden of decades of high-intensity trekking tourism in an ecosystem as fragile economically as it is environmentally. Know what to look for in a group trek, understand environmental impacts, and consider volunteering time and skills to local communities along your trek.
HOW Research porters' rights (www.ippg.net) and environmental sustainability (ntnc.org. np) before you book.

© DAVID DORAN

TEST YOUR METTLE

Nothing cements a travel experience in your memory more than a little bit of 'Type 2' fun.

For many travellers, the idea of going to a destination to relax is a complete anathema. Far from hotel pools, beach loungers and all-inclusive resorts, there's a wild world to be explored, and one of the best ways to really see a new place is through physical activity. And a bit of sweaty suffering. Whether you're biking, hiking or skiing through the hills, skating along a frozen river or wild swimming around a bit of craggy coastline, the experience might be excruciatingly exhausting at the time, but so long as you get through it in one piece, the rewards are plentiful – especially in retrospect, when your muscles have forgiven you and your brain has forgotten the pain.

One immersive option is to base your entire trip around a specific physical challenge – such as completing a peculiarly punishing long-distance walking trail or summiting a particular peak, such as Mont Blanc. Lots of runners decide their travel plans around iconic races, for example the Paris, Boston and New York marathons, while others might want to go further and experience conditions in the highlands of Kenya and Ethiopia, which have produced so many great champions. Cyclists like to almost break themselves riding on the cobbles of Flanders and up the Alpine passes that they've watched their heroes battle, and paddlers can take on numerous gnarly canoe and kayak routes.

BIRKEBEINERRENNET
Norway

A 33.5-mile (54km) classic cross-country ski race from Rena to Lillehammer, this event commemorates a journey undertaken in the 13th century, when two men saved an infant prince by smuggling him through these mountains. To represent the baby, all participants must carry a backpack weighing at least 7.7lb (3.5kg).
HOW Sign up at birkebeiner.no.

COMRADES MARATHON
South Africa

If your standard marathon sounds too easy, try the 55-mile (89km) Comrades road race between Durban and Pietermaritzburg. There's a strict cut-off time (11 or 12 hours) and a spirit of *ubuntu* (a Nguni Bantu term meaning 'humanity') prevails. Held in June.
HOW The field is limited to 25,000 and it sells out. See www.comrades.com.

HELLESPONT AND DARDANELLES SWIM
Turkey

This annual open-water cross-continental challenge closes one of the world's busiest shipping lanes as people swim 2.8 miles (4.5km) from Eceabat to Çanakkale (Europe to Asia), across the Dardanelles. Held in August.
HOW SwimTrek (www.swimtrek.com) organises four-day packages.

YUKON RIVER QUEST
Canada

The world's longest annual river race sees canoeists and kayakers paddle 444 miles (715km) along the Yukon River under the midnight sun, as prospectors did during the Klondike Gold Rush in the 1890s. Held in June.
HOW Part of the compulsory kit list is bear spray. Sign up at www. yukonriverquest.com.

Racy stories

Anyone can have a crack at a marathon – but could you beat a horse in a running race? That not-quite-million-dollar question was posed in a pub one night in Wales in 1980, which led to a bet, which in turn led to one of the world's wackiest races: Man versus Horse. Each year, in Llanwrtyd Wells, 650 runners race 60 horses (with riders), across the same 22-mile mountainous course. A £500 prize was put up for the first human to beat the fastest horse, and it was increased by £500 each year until Huw Lobb achieved the fleet-footed feat in 2004, pocketing £25,000. Three years later, Florien Holtinger also won – but the prize has remained unclaimed ever since. Fancy giving it a go? Enter at www.green-events.co.uk.

BUILD UP TO IT
Aim for the top:
Summit a mountain, p152
Increase your stamina:
Come home fitter, p166

HELP SAVE AN ENDANGERED SPECIES

Pay it forward to the world's most vulnerable inhabitants by signing up for a volunteer project or supporting a tourism initiative that helps to ensure their survival. You'll gain an insight into the human impact on their world and learn how to mitigate yours.

*E*ven the best nature documentaries don't compare with the visceral thrill of observing a rare animal in the wild. By opting to take part in a wildlife volunteer project on your trip, you can play an active role in ensuring your favourite critters will be around for the next generation of travellers to marvel at, as well as pick up new skills such as learning to dive or how to treat injured animals.

There are thousands of wildlife volunteer projects to choose from around the world, from caring for rescued elephants in Thailand to monitoring coral reefs in Mexico. Before you dive into a placement, do the research to ensure the project will legitimately help to protect the animals you'll be working with, and matches your skills and expectations. A good starting point is Responsible Travel (www.responsibletravel. com), which handpicks responsible wildlife volunteering options from 400 specialist companies worldwide. No matter which company you choose to support, don't be shy to ask questions about how the operation is run.

Short on time? By opting for a hotel or organised tour that contributes a percentage of profits to wildlife protection projects, you can still do your bit to help save a species. Entry fees for wildlife sanctuaries also tend to be critical for keeping their conservation programmes running, so you can feel good about visiting them, too.

KARONGWE PRIVATE GAME RESERVE
South Africa

Rhino poaching is still at crisis levels across Africa, so why not lend your support to an anti-poaching initiative? GVI offers a volunteer placement in South Africa's Karongwe Private Game Reserve focused on creating increased awareness around rhino poaching. During a minimum two-week placement, volunteers learn how to track wildlife, collect vital data and get involved in a wide range of conservation activities. **HOW The Rhino Poaching Awareness programme (www.gvi.co.uk) includes basic accommodation, meals and training.**

INKATERRA MACHU PICCHU PUEBLO HOTEL
Peru

Optioning a trip to Peru? Splash out on a night at the Inkaterra Machu Picchu Pueblo Hotel in the village below the world's most famous Inca ruin. This hotel funds its own sanctuary for vulnerable Andean (also known as spectacled or 'Paddington') bears. For a small donation, guests can opt to tour the excellent sanctuary designed to recover bears that have been negatively affected by humans, such as bears kept as pets. **HOW The Inkaterra hotel (www.inkaterra.com) is located in Aguas Calientes, the accommodation hub for Machu Picchu. It's almost two hours by train from the Sacred Valley town of Ollantaytambo, 45 miles (72km) northwest of Cusco.**

ILE AUX AIGRETTES
Mauritius

Take a day off from lazing on the beach in Mauritius to be a 'conservationist for a day' on Île aux Aigrettes, a nature reserve just a short hop from the mainland. The experience – which might include preparing breakfast for baby giant tortoises – supports the Mauritian Wildlife Foundation's work protecting the island's endemic species including the critically endangered olive white-eye songbird. **HOW This experience, led by a Mauritian Wildlife Foundation guide, is bookable through Mauritius Conscious (www.mauritiusconscious. com), the country's first dedicated sustainable travel agency.**

Left: Recently hatched baby turtles

© PHILIP LEE HARVEY | LONELY PLANET

Sea turtles: a conservation success story

Offered everywhere from Thailand to Costa Rica, sea turtle conservation is one of the world's most popular wildlife volunteering projects, and there's now scientific proof that it's working. Despite six of the seven variants of sea turtle being endangered due to humans killing them for their eggs, meat and shells, a 2017 study published in the Science Advances (www. advances.sciencemag.org) journal found that numbers of all seven species are increasing, with the research team from Australia and Greece attributing the positive trend to the effective protection of eggs and nesting females, as well as reduced by-catch. The findings highlight the importance of continued conservation efforts that underpin this global conservation success story.

BUILD UP TO IT
Take pressure off the well-trodden path:
Get off the tourist trail, p64
Learn better habits:
Travel without plastic, p204

STAND AT THE END OF

Only certain, far-flung places have the vibe: frothing sea, brooding sky, unrelenting wind, stunted trees. The road ends, and there's nothing more. You stare over the edge, a speck admiring the vastness, and then it hits you: the clarity that comes from feeling so very small.

THE WORLD

USHUAIA
Argentina
Set at the desolate tip of South America where the Andes dump into the sea, Ushuaia is officially dubbed The End of the World, with a much-photographed sign to prove it. Its shipwreck-strewn waters are the last call before Antarctica.
HOW Tierra (www.tierradelfuego.org.ar) offers hiking, kayaking and skiing tours.

FOGO ISLAND,
Newfoundland, Canada
Fogo floats by its lonesome off Newfoundland's coast, facing nothing but bleak northern seas. The Flat Earth Society claims it's one of the four corners of the globe, so you really are at the edge while watching the icebergs drift by.
HOW The closest airport is in Gander. From there, drive an hour north to Farewell to catch the hour-long ferry to Fogo. www.townoffogoisland.ca.

NORDKAPP (NORTH CAPE)
Norway
On the tip of Norway's northern coast, Nordkapp is spitting distance from the North Pole, and hauntingly beautiful. It's the view that thrills the most: gaze down at the wild surf and absorb the immensity.
HOW The nearest airports are in Alta and Lakselv; Nordkapp is a 3.5-hour drive north from either. See www.nordkapp.no.

CAPE AGULHAS
South Africa
Agulhas is the southernmost tip of Africa, where the Atlantic and Indian Oceans collide. The wind-bashed coastline is the graveyard for many a ship. A forlorn, red-striped lighthouse provides views of the scene.
HOW The closest airport is in Cape Town. From there, rent a car and head 2.5 hours southeast. See https://discovercapeagulhas.co.za.

BUILD UP TO IT
See what others left behind:
Explore an abandoned place, p16
Relish the throng:
Witness a swarm, p112

© RAPHAEL KOERICH | 500PX | GETTY IMAGES

Left: Tierra del Fuego, Ushuaia

Veganism, vegetarianism and other plant-based diets are no longer food lifestyles on the fringe. There are more carnivores in transition than ever before, and for moral, environmental, economic and gustatory reasons. Going meat-free during a trip is a great way to road-test these new preferences.

Even committed omnivores acknowledge that plant-based food diets are more sustainable than those incorporating meat. The economic and environmental costs of raising animals for slaughter are, pound for pound, much higher than growing grains and pulses, especially when factoring in the feed stock (and pesticides), water and space required for animals, and the resulting waste and greenhouse gas emissions.

So what should the rank and file do to reduce their agricultural footprint? Cut back on eating meat, for one. And why not use the disruption caused by travel as a test run, especially for flexitarians (who sometimes eat meat) and reducetarians (who aim to curb their meat consumption)? If nothing else, it's less expensive.

Fortunately, many countries are distinctly vegetarian-friendly, which makes life easier. Asia claims quite a few: India, Vietnam, Thailand and Sri Lanka, for example. Elsewhere, there's Italy in Europe and Ethiopia in Africa as well as food-forward countries that are increasingly embracing meat-free practices, such as the USA, UK and Germany.

There are also resources to help direct the blossoming demand for plant-based food tourism. VeggieHotels (www.veggie-hotels.com) and VeganWelcome (www.vegan-welcome.com) simplify finding the right places to stay, while HappyCow (www.happycow.net) has listings for veg-friendly, vegan, raw and some organic cafes and restaurants.

GO MEAT-FREE

COM CHAY RESTAURANTS
Vietnam

In Vietnam, a 'Com Chay' sign identifies vegetarian eateries in which any meat-like ingredients are tofu- or soy-based. Often these are local, simple places popular with observant Buddhists. Vegans and vegetarians should be alert to the widespread use of fish sauce and shrimp paste; meat stock fills most pho, even when the added extras are veg only.

HOW *Com Chay* restaurants are found throughout Vietnam, often adjacent to Buddhist temples.

RICE AND CURRY
Sri Lanka

Sri Lanka has some of the world's most varied, delicious and spicy vegetarian food, heavily influenced by Buddhist and Hindu beliefs, as well as ayurveda. Think dhals, vegetable curries, coconut salads, papadums, rice and more. Vegans should beware of ghee (clarified butter), cheese and coconut powders that include dairy.

HOW In Negombo, not far from the international airport, Jetwing Ayurveda Pavilions (www.jetwinghotels.com) is a stunning vegetarian hotel.

PASTA
Italy

Most fresh Italian pastas use only three ingredients: water, flour and egg. When combined with cheese or vegetable sauces, they deliver rich pickings for vegetarians. Vegans should avoid dairy-based sauces and ask for any pasta *senza uovo* (without egg); veg-only antipasti and pizza (the typical dough is vegan) are also flavourful fall-backs.

HOW Intrepid Travel's line of vegan tours includes an eight-day food adventure in Italy. www.intrepidtravel.com.

ON THE ROAD

BUILD UP TO IT
Sample new things:
Eat like a local, p38
Live in harmony with your surroundings:
Go off grid, p192

EXPERIENCE COMMUNITY LIVING

Technology is said to have made the world smaller. People are more hyper-connected than ever. But, counter-intuitively, many of us are also lonelier and increasingly searching for meaningful human interactions. Travel makes that possible, especially when you spend time in communities where it's all for one and one for all.

*U*nderpinning much of human history is a search for community – finding or creating a sense of belonging in the company of others. Is it any wonder then that immersive travel embraces community living? It's definitely worth giving it a shot, whether the desire is for spiritual exploration in a monastery or ashram, or solidarity through shared projects such as in a self-sufficient eco community. More recently, in the age of gig-based freelance employment, physical isolation has spurred new types of togetherness in networks of co-working and co-living spaces.

Unifying the members of most communities is a readiness to lend a hand in group pursuits that serve the greater good, usually in exchange for room, board and, of course, camaraderie. Spending time in a shared community is not always easy – tension and arguments about the washing-up seem to happen wherever humans co-exist – but the experience can also be deeply affecting, cementing values and friendships that can last a lifetime.

SUNSEED DESERT TECHNOLOGY
Spain
The lively, international community at Sunseed Desert Technology, near Almería, has been involved in organic farm work and research into sustainable living practices for 30 years. Visitors are expected to participate for a minimum of four hours a day.
HOW Sunseed Desert (www.sunseed.org.uk) accepts volunteers for stays of up to seven weeks. For other similar projects, see www.ecovillage.org.

AMRITAPURI ASHRAM
India
Mata Amritanandamayi Math, the ashram's full name, is the Kerala birthplace and headquarters of Amma, the 'Hugging Saint'. All are welcome and participate in *seva* (selfless service) – from washing to serving food – to keep things running.
HOW Visitors must register in advance (amma.org) and abide by the ashram's spirit and practices, guided by Amma's teachings.

UNSETTLED
Worldwide
Unsettled runs retreats for digital nomads and creative entrepreneurs looking for like-minded community support and a co-working space. There is a flexible schedule of activities and workshops, and members are encouraged to give something back to the group, whether that's through mentoring, sharing skills or leading a meditation class.
HOW Retreats are held for two weeks or one month in select destinations around the world. Find out more at www.beunsettled.co.

Kibbutzim

The kibbutz movement began in Israel in 1910, attracting idealist volunteers yearning to get back to nature. Today, after an exodus and then a return of the kibbutz youth, there are 270 kibbutzim (the plural of 'kibbutz') and many of these projects have morphed and modernised.

If you're between 18 and 35, it's possible to spend two to six months on a traditional kibbutz. Volunteers will help with manual labour, which could include anything from washing up to milking cows. Food and accommodation are provided.

For short-term travellers, modernised kibbutzim are also opening their doors to outsiders. This offers the chance to partake in classic kibbutz traditions, such as the *hadar haochel*, communal dining, while spending the rest of your time relaxing at a spa or hiking trails through aromatic mango groves. Ein Gedi Kibbutz Hotel (www.ein-gedi.co.il) on the shore of the Dead Sea, and Kibbutz Lotan Desert Inn (www,kibbutzlotan.com) 40 minutes from Eileat both welcome day guests and longer term visitors. For more information on volunteering on a kibbutz, visit www.kibbutz.org.il/eng.

BUILD UP TO IT
Make new friends:
Find your tribe in a foreign city, p90
Avoid the mainstream:
Live on the fringe, p190

© UNSETTLED

Left: Community life at an Unsettled retreat

TAKE A BIG TRIP ALONE

No longer as daunting as it once seemed, solo travel is on the up: be it on a multiday mountain hike, city break or yoga retreat. Strike out alone and you're the master of your time, budget and destiny, with no need to compromise.

oing solo is trending. ABTA's Holiday Habits 2018 report showed that one in nine travellers now holiday alone, double the figures of six years earlier. As the world opens up, lone travellers are no longer willing to wait for the perfect companion before kick-starting once-in-lifetime adventures. Instagram pics of dreamy destinations inspire, and a raft of handy apps make travel a breeze. Some still cite the dangers, but pick the right destination and solo travel can be liberating: boosting self-awareness and confidence, making you more spontaneous and letting you control the itinerary. Want to crash on the beach all day? Spend an afternoon nosing around bookshops? Finally begin your novel? Blow your budget on one fancy meal? Go right ahead.

Quick to embrace the momentum, there are now a raft of trips out there if you would rather join up with a group: whether you want to go hiking in Bhutan, skiing in the Alps or on an expedition cruise to Antarctica. But you yourself can easily find your own like-minded crowd: surfing breaks in Portugal, perfume-making workshops in Provence, wine tours in Australia, yoga retreats in India.

Forcing yourself out of your comfort zone often means your trip will be as much about self-discovery as anything else. Travelling alone also makes you more sociable and approachable, more likely to connect with locals and fellow travellers, and perhaps give the language a go. So, in that sense, travelling alone really isn't lonely in the slightest.

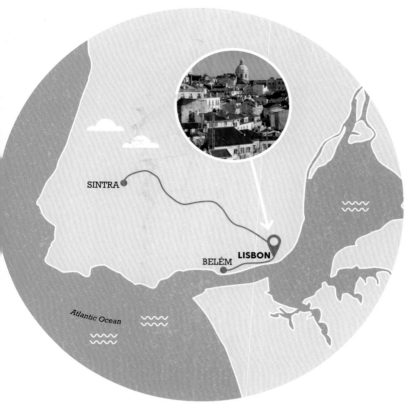

FOUR DAYS: LISBON & AROUND

Portugal's petite, high-spirited capital, right on a river opening out to sea, wins our solo city-break vote.

Spend a couple of days exploring the picturesque lanes of the Alfama district and pedestrianised Baixa, including a must-do ride on tram 28E. Then take a day trip to **Sintra** to wander through the boulder-speckled woodlands and fairy-tale palaces. Back in Lisbon spend your final day in **Belém**. Stroll the waterfront and explore the excellent museums and fantastical Mosteiro dos Jerónimos. Say goodbye to the city with a night out in rowdy Cais do Sodré.

SOLO TRAVEL TIP Visit spring through autumn, avoiding August if you can. Opt for an apartment rental or small guesthouse in a sociable, walkable neighbourhood such as Alfama or Graça, where you'll get to know the locals.

TWO WEEKS: VIETNAM

The adventure begins in the cauldron of commerce that is **Ho Chi Minh City**. Spend two to three days hitting the markets, browsing museums and eating some of Asia's best cuisine.

Then it's a plane or train up to **Danang** to access the cultured charmer and culinary hot spot that is **Hoi An**. This town certainly warrants three or four days, such is its allure. Enjoy Hoi An's unique ambience, touring its temples and Old Town, and visit the nearby beach of An Bang. Then it's on to the old imperial capital of **Hue** for two to three nights to explore the citadel, pagodas and tombs (and nearby beaches, in season).

Next, it's a long journey by train (or a flight) to **Hanoi** to check out the capital's evocative Old Quarter, munch some street food and view the city's elegant architecture and cultural sights. From Hanoi, book a tour to incomparable **Halong Bay**, which boasts more than 2000 limestone islands, before returning to Hanoi.

SOLO TRAVEL TIP Avoid rip-offs by asking your accommodation staff how much things should cost (such as taxi rides to tourist sights) before you head out for the day.

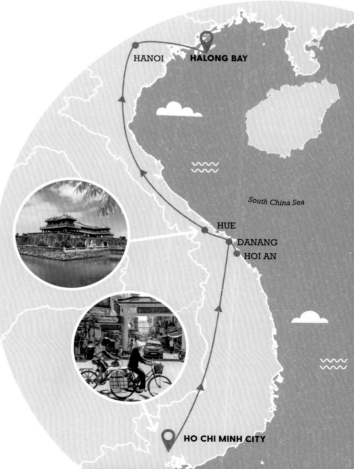

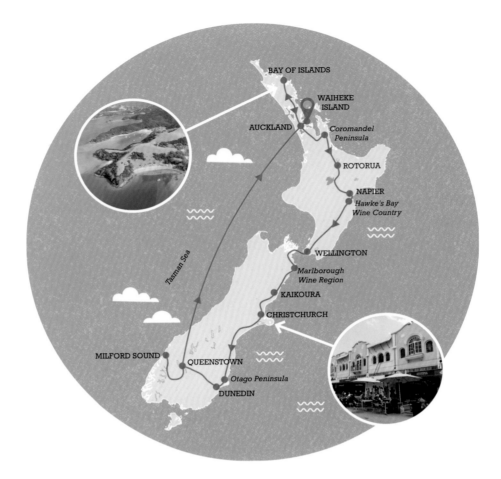

ONE MONTH: NEW ZEALAND

Classy cities, geothermal eruptions, fantastic wine, Māori culture, glaciers, extreme activities, isolated beaches and forests: just a few of our favourite NZ things.

Aka the 'City of Sails', **Auckland** is a South Pacific melting pot. Spend a few days shopping, eating and drinking: this is NZ at its most cosmopolitan. Make sure you get out on to the harbour on a ferry or a yacht, and find a day to explore the beaches and wineries on **Waiheke Island**. Truck north to the **Bay of Islands** for a dose of aquatic adventure (dolphins, sailing, sunning yourself on deck), then scoot back southeast to check out the forests and holiday beaches on the **Coromandel Peninsula**. Further south in **Rotorua**, get a noseful of eggy gas, confront a 33ft (10m) geyser, giggle at volcanic mud bubbles and experience a Māori cultural performance.

Meander down to **Napier** on the East Coast, NZ's attractive art-deco city. While you're here, don't miss the bottled offerings of **Hawke's Bay Wine Country** (...ohh, the chardonnay). Down in **Wellington**, the flat whites are hot, the beer's cold and wind from the politicians generates its own low-pressure system. This is New Zealand's arts capital: catch a live band, buskers, a gallery opening or some theatre, then take a hike up Mt Victoria for the city's most impressive viewpoint

Swan over to the South Island for a couple of weeks to experience the best the south has to offer. Start with a tour through the sauvignon blanc heartland of the **Marlborough Wine Region**, then chill for a few days between the mountains and the whales offshore in laid-back **Kaikoura**. Next stop is the southern capital **Christchurch**, swiftly rebuilding after the earthquakes. Follow the coast road south to the wildlife-rich **Otago Peninsula**, jutting abstractly away from the Victorian facades of Scottish-flavoured and student-filled **Dunedin**. Catch some live music while you're in town. Head inland via SH8 to bungy- and ski-obsessed **Queenstown**. Don't miss a detour over to **Fiordland** for a jaw-dropping road trip and boat cruise around **Milford Sound**, before returning to **Queenstown** for your flight back to **Auckland**.

SOLO TRAVEL TIP Be aware that booked activities are often cancelled for weather reasons. If you have your heart set on a helicopter ride or wildlife walk that could be rained off, build extra time into your itinerary in case your tour is bumped to the following day.

SIX WEEKS: SOUTH AMERICA

This classic South American journey takes in some of the continent's most famous sites, including Andean peaks, Amazonian rainforest, Iguazú Falls and the inimitable Galápagos Islands.

Start off in **Buenos Aires.** Spend several days exploring the mesmerising Argentine capital. Go west to **Bariloche** for spectacular scenery then head to Chile's verdant Lake District at **Puerto Varas.** Continue north to **Santiago**, then cross back into Argentina to **Córdoba** and gorgeous **Salta** before re-entering Chile at the desert oasis of **San Pedro de Atacama.** Head into Bolivia to experience the surreal **Salar de Uyuni.** Continue to **La Paz** and on to Peru via **Lake Titicaca.** Linger at ancient Cuzco before going to Lima and on to Ecuador.

From **Guayaquil**, fly to the Galápagos Islands. Back on the mainland, visit colonial **Cuenca** and historic **Quito.** Pass into Colombia to the lush Zona Cafetera and bustling **Medellín**, then go to Cartagena for Caribbean allure. See beautiful **Parque Nacional Natural Tayrona**, then head towards **Leticia** and cross into Brazil. Head to **Manaus** for a jungle trip. Afterwards fly down to **Rio de Janeiro** for beaches and nightlife. Visit thundering **Iguazú Falls** and return to **Buenos Aires.**

SOLO TRAVEL TIP Consider taking preventative medication to thwart altitude sickness, which is a risk in Bolivia.

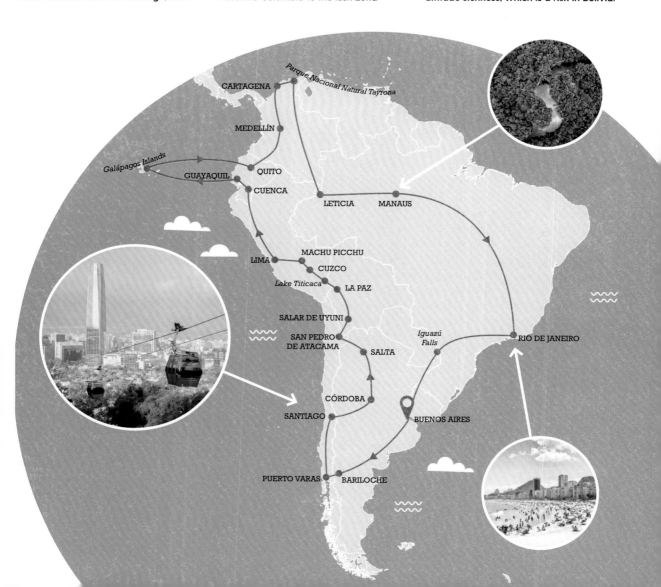

LEARN ABOUT FRAGILE PLACES

These are turbulent times for our planet. Terra doesn't feel too firma. Facing a choice, as we humans do today, between striving for greater natural equilibrium or tilting towards irreparable imbalance, learning about what it takes to be better earth stewards might help.

Is there any denying the beauty of our solitary blue planet – the sensory-rich mix of biodiverse ecosystems teeming with millions of living species? It's a gorgeous home.

But it's also our only home, which is increasingly worrisome as, despite its hardiness, the earth shows clear signs of distress. The demands of growing human populations, especially the effects of industrial and technological change, have dealt serious blows to the health of the environment. Particularly alarming is the speed with which the climate is changing, creating weather extremes that are altering those glorious ecosystems: glaciers are melting, water-starved forests are tinder for infernos and oceans are warming with serious deleterious consequences to all life.

Here's the thing: it's not too late. Not too late to travel (responsibly!) to places where the fragility of our home is in evidence. Not too late to learn about and join efforts to correct course in those places, and influence others to do the same. Because seeing really is believing and action is what matters, especially when it leads to beneficial behaviours like more personal accountability, sustainable practices, protection for and enhancements to the natural world, and green thinking and design.

RESTORE CORAL REEFS

The Caribbean

Reef Renewal Foundation Bonaire is one of many organisations in the Caribbean and the Pacific working furiously to establish coral nurseries, grow new coral and plant new reefs. Visiting divers help to build previously expansive fields of coral back into complex and healthy marine habitats.

HOW Bonaire is a year-round dive destination. Participation in coral restoration requires specialised dive instruction, which is available locally. See www.reefrenewalbonaire.org.

WITNESS GLACIAL RETREAT

Alberta, Canada

The Athabasca Glacier, one of the six main tongues of Canada's enormous Columbia Icefield, is the most visited glacier in North America. For the adventurous, it's easily reached via giant, all-terrain Ice Explorers. The nearby Glacier Discovery Centre describes the life of a glacier, as well as the Athabasca's dramatic retreat.

HOW Outside the winter closure periods (November to March), access is from the Icefields Parkway in Alberta's Jasper National Park. www.banffjasper collection.com.

CONFRONT NEW FOREST FIRE REALITIES

USA

In recent years, deadly fires have ravaged forests on all continents except Antarctica, devastating natural areas, and displacing people and animals. Climate scientists describe the fires as part of the new normal. In many US national parks with forests in recovery, signs put the fire causes and effects into context.

HOW The destruction is dramatic in US parks such as Yellowstone and Yosemite. See www.nps. gov/subjects/fire.

Left: Coral and marine life in Bonaire

Reef renewal in Bonaire

Bridget Hickey is the assistant coordinator at Reef Renewal Foundation Bonaire, which works to preserve Bonaire's coral reefs through protection and restoration. 'We've planted over 20,000 corals on the reefs of Bonaire,' she says. 'To do that, we've relied heavily on local dive operators and people who want to give something back. The latter spend a day and a half getting trained by dive-shop instructors who specialise in what we do: clipping corals in nurseries, tying them up in underwater trees and planting them on reefs. They make a real impact in devastated areas where once there was no life, no fish, almost nothing.'

BUILD UP TO IT

Make a difference to what's around you:
Volunteer at home, p74
Take your foot off the gas and put it on the pedal:
Travel by bike, p170

EXPERIENCE A WEEK OF SILENCE

STEP FOOT ON THE FROZEN CONTINENT

FACE YOUR FEARS

MASTER A FOREIGN TONGUE

WITNESS A MIR

A MI

OFF THE LAND

HELP A COMMUNITY REBUILD

MAKE AN EPIC OVERLAND JOURNEY

BE YOURSELF

ACLE OF NATURE

MEET YOUR HERO

DON'T STOP TRAVELLING

6

STRIP OFF

Shed your clothes and your inhibitions by breaking one of the most common global taboos: nudity. It can help you come to terms with the kind of body image issues a lot of us face. Plus, feeling the wind in unusual spots is always a unique kind of treat.

Maybe you're the kind of person who changes in the bathroom stall at the gym, still traumatised by memories of group showers in high school phys ed. Or maybe you're already a 'naked person', happily striding around your (curtain-less) house in the altogether. Either way, unless you're a committed 'naturist' (as those once referred to as 'nudists' prefer to be called), you've probably never partaken in the activities we're about to suggest here. Things like nude skiing, co-ed bathing sans swimsuit, hiking through the forest with only a sunhat, even eating dinner in a restaurant completely starkers.

'Isn't the invention of clothing one of the finest human achievements?' you may say. Sure, and so are cars and airplanes, but it's still good to walk around on your feet. Being naked connects us with a more primordial version of our self, a version without all the body hang-ups and self-conscious style worries. Clothes can become emotional armour, as any expensively black-clad city-dweller will tell you. Taking off that armour can let us be freer and more open.

Also, come on, the nude skiing will make an *excellent* story to scandalise your grandkids with some day.

NUDE HIKING
Germany
Germany is generally pretty tolerant of nudity, but this 11-mile (18km) trail is a first. Running through the Harz Mountains between the towns of Dankerode and the Wippra Dam, the Harz Naturist Climb is popular with the local clothes-free community. Trails are marked with signs warning clothed hikers of what to expect. Bring decent hiking shoes, plenty of sunscreen and a backpack for your clothes.
HOW Camping Panoramablick in Dankerode offers campsites as well as a nude beach. The nearest city is Leipzig, a 90-minute drive away by car.

SUIT-FREE SNORKELLING
St-Martin & Sint Maarten
This half-Dutch, half-French Caribbean island has long been a nude-friendly place. You can lounge and swim nude on Orient Beach, or hop aboard an in-the-buff boat tour that includes stops for snorkelling and sunning in secluded coves. Feeling the warm water rushing over your body as you glide over the reef is, quite frankly, delightful. Ditto the sensation of sun and wind on your back as the speedboat bounces over the waves.
HOW Day-long tours depart on Tuesday and Thursday from Oyster Pond Marina; see www. sxmdeals.com/st-maarten-excursions-nude-clothing-optional-tours. Stay at Esmeralda Resort (www. esmeralda-resort.com) on nude-friendly Orient Bay.

NAKED LUNCH
Paris
This 12th-arrondissement establishment looks like any other discreet modern bistro: black chairs, sophisticated blue walls, contemporary pale-wood tables. Only O'naturel is a leeetle more discreet than most: it has no windows. Opened in 2017, it serves classic Parisian fare to an all-nude clientele (waiters and waitresses are clothed). Dine on foie gras and pears in caramel. Change in the locker room and leave your phone behind.
HOW The restaurant opens 7.30pm to 11pm Monday to Saturday. Reservations are required. See www. restaurant-onaturel.fr. 9 Rue de Gravelle, Paris.

Left: St-Martin & Sint Maarten in the Caribbean is popular with naturists

What is a naturist?

When you think about naturists, you may be imagining middle-aged couples playing volleyball at some mid-Atlantic cabin resort. But according to The Naturist Society (www.naturistsociety. com), its members are all ages and from all socioeconomic classes: single people, couples, families, teachers, computer programmers, retirees. They all share the same underlying principle: 'body acceptance through nude recreation' of all sorts, from hiking to dancing to cooking. The society views the nude body as 'a gift of nature, dignified and worthy of respect, regardless of shape, size, age or hue'. Make sure you partake in nudity in designated or accepted areas and being naked becomes comfortable, liberating and fun. It's not about sex at all, they say: 'nude is not lewd' is a popular naturist mantra.

BUILD UP TO IT
Don't take it so seriously:
Travel for laughs, p24
Feel good in your own skin:
Treat your body, p50

A voyage to the end of the world as we know it might well be the missing link in understanding our planet and the need to protect its wild places. Brace yourself for an awakening in an off-the-radar continent that blows all conventional notions of 'beauty' out of the water.

Antarctica is more than just another 'trip-of-a-lifetime' to tick off a bucket list: this is a journey that is likely to bring about moments of enlightenment that challenge the way you perceive the planet and your place in it. The White Continent has a singular beauty. Here, opalescent-blue icebergs tower like the epic ruins of fantasy castles, glaciers calve with a colossal roar to send penguins in their hundreds skittering across snowy beaches, and whales of every species breach in glass-clear waters.

Getting here by expedition cruise on an ice-strengthened polar vessel isn't something to be take lightly: you have to earn Antarctica. The bone-chillingly cold up on deck, the lonesome skies, the relentless swaying on the wave-lashed, 600-mile-wide, two-

day Drake crossing – it's all part of it. Then, finally, you emerge in still waters that reflect a light too brilliant for this world. In these waters, mountains rise up like daggers, penguins, porpoise, humpbacks blow and tail-slap, and fur and leopard seals drift by on icebergs within a hair's breadth of the boat. This is your reward.

In Antarctica, one thing quickly becomes clear: nature is in control and humans are just passing through. And that thought, well, it changes you. All your senses are on high alert and in-the-moment: feeling the cold, listening to the honking of penguins, watching the ice break. This is perhaps as close as we can get to the eternal; the continent where nature has its finest moment, time seemingly stands still and reality takes on a whole new dimension.

STEP FOOT ON THE

MARINE MAMMALS EXPEDITION

Departing from Ushuaia at the southernmost tip of Argentina, a marine mammals expedition is a great way to dip your toes into Antarctica. The Antarctica cruise season is from November through March, the latter being the prime time to see marine mammals. Small, conservation-focused polar vessels, with scientists on board and daily zodiac trips, are the way to go. Sharing a cabin can slash costs.
HOW One Ocean Expeditions (www.oneoceanexpeditions. com) offers 10-night marine mammals cruises in March, including meals, equipment and activities.

LONGER AND SPECIALIST VOYAGES

With more time (and money) on your hands, Antarctica is your oyster. Options include photography-focused cruises, offbeat expeditions with hiking, kayaking and snowshoeing, and multi-week journeys taking in the Antarctic Peninsula, South Shetland Islands and wildlife-rich Falkland Islands.
HOW In November One Ocean Expeditions (www. oneoceanexpeditions.com) offers a 12-night package, with activities such as sea kayaking and ski touring.

FLY-CRUISE OPTIONS

If time is of the essence, you can dodge the notorious two-day Drake Passage and instead fly directly to Antarctica in just two hours – from Punta Arenas in Chile to King George Island in the South Shetlands.
HOW Swoop Antarctica (swoop-antarctica.com) can arrange the logistics, with departures from early December to February. Fly-cruise trips tend to be slightly pricier than an expedition voyage.

Below: Emperor penguins on the Dawson-Lambton Glacier, Antarctica

ROZEN CONTINENT

BUILD UP TO IT
Wake up in a different country:
Take the sleeper train, p20
Travel for the visual spectacle:
Be uplifted by colour, p72

PULL TOGETHER FOR THE TEAM

Get a gang of mates together and turn them into wannabe world-beaters by taking on a challenge in a far-flung place.

mbracing adventurous challenges while travelling is a great way to see the wider, wilder world, but doing so as part of a posse is even better, as it forges lifelong friendships and forms shared memories of experiences in exciting, exotic places. Either that or you end up never speaking to one another again... but, hey, what a great way to test your compatibility. Whether you're pulling together as a team to take on a relay challenge, pooling talent to ascend a mountain top or learning the ropes as a brand-new crew in order to sail a yacht, a sense of camaraderie emerges surprisingly quickly when you are pitting your collective wits against the elements (and other teams) in pursuit of a common goal. What's more, when you nail that objective or

even just manage to reach the finish line together, the collective euphoria you'll feel will be off the scale.

Team experiences can be carefully preplanned – you and a buddy might enter a clapped-out car rally across the Mongolian steppe – or you can leave it to serendipity and sign up for a group activity on your lonesome (perhaps a stint working on a tall ship as it makes an ocean crossing) and see what fortune throws your way in terms of who you meet and bond with. One excellent way to organise a team-based travelling experience is to get a mob of like-minded mates together and enter an international adventure race – between the pre-travel training and the event itself, you'll soon see whether you're really cut out to be comrades.

BUILD UP TO IT
Appreciate a small community:
Become absorbed in village life, p28
Feel welcome:
Find your tribe in a foreign city, p90

Left: Sailing in the Adriatic

24-HOUR PARTY PEDALLERS
Scotland and Canada
Cycling through the night isn't everyone's cup of chain lube, but entering a 24-hour mountain-bike enduro as a team changes the dynamic of the experience, as you alternate between riding laps and cheering on your mates. At many muddy mash-ups – like the Strathpuffer (which takes place in the midst of a Scottish winter) and the 24 Hours of Light in Whitehorse, Canada (where the sun never sets) – a festival-like atmosphere prevails around the event hub.
HOW See www.strathpuffer.co.uk and www.24hoursoflight.ca for more information.

BUILD A RAFT AND RACE IT
Australia and Peru
Harness your inner Huck Finn, build a boat and race it along a wild waterway. Not for the faint-hearted, the 20-year-old River Amazon International Raft in Iquitos, Peru, involves four-person teams taking on 118 miles (190km) of the Amazon River over three days in self-made rafts. More accessible, Australia's Darwin Beer Can Regatta sees four-person teams construct craft from, yup, beer cans, and race them on Darwin Harbour.
HOW The 'Ten Can-mandment' decree all boats must be made from cans, but juniors can use soft drink cans; see www.beercanregatta.org.au.

THE THREE PEAKS YACHT RACE
UK
One of the planet's oldest and toughest multisport challenges, this race sees teams of sailors, cyclists and runners working together, harnessing human power and the elements, to ascend Wales, England and Scotland's highest peaks. Starting in Barmouth and finishing in Fort William, two runners from each team must summit Snowdon, Scafell Pyke and Ben Nevis, with the boat-based crew sailing them between each. Teams can cycle to the bottom of Scafell, but no engine power is permitted.
HOW See www.threepeaksyachtrace.co.uk.

MEET YOUR HERO

It's wild who you'll run into while globetrotting, whether by design or by chance. If you're lucky, travel might put you in your idol's orbit where unforgettable meetings can result.

It happens more than you think. You meet Robert Redford in the Los Angeles airport lounge, or your favourite author on the train from London to Paris. The encounters seem like pure luck, but you helped them occur by travelling – you were in the right place at the right time.

Of course, you can also travel specifically to seek out your hero. Journey to Milan to meet your favourite bicycle designer at his shop, for example. Visit Lagos to go to the adored Afrobeat singer's club. Make the trip to India to greet your guru. Even if your heroes are

no longer alive, you might be able to tour their home or studio, see where they ate their lunch or played with their children, and 'meet' them in spirit, which is inspirational in its own right.

Remember, we're not talking about stalking VIPs or bothering them if they want to be left alone. And there's the old adage about never meeting your heroes because you'll find out they have feet of clay. That could happen. But often serendipity is on your side, and when you bump into someone you've admired from afar, he or she turns out to be all you expected and then some.

SANTA'S PUB
Nashville, USA
Kid Rock, Mumford & Sons, the Lumineers and many more musicians from the local scene pop into this dive bar set in a double-wide trailer. The white-bearded bartender, year-round Christmas decorations and democratic karaoke are the draws.
HOW Mingle with your rockin' heroes at Santa's Pub (www.santaspub.com) from 4pm to 2.30am daily; karaoke starts at 7pm (9pm Sunday).

BAEKYANGSA TEMPLE
South Korea
Jeong Kwan is a Zen Buddhist monk and star chef admired for her exquisite vegan cooking that was featured in the Netflix documentary *Chef's Table*. She lives and works at Baekyangsa Temple, which is located about 170 miles (274km) south of Seoul.
HOW Stay at the temple (www.eng.temple stay.com) and try Jeong Kwan's dishes.

Below (left and right): Frida Kahlo's Casa Azul (Blue House) is now an art museum dedicated to her life and work

BEATLES' CHILDHOOD HOMES
Liverpool, UK
The National Trust offers tours of the boyhood homes of Paul McCartney and John Lennon. Paul's place, in particular, feels like the real deal, complete with threadbare sofas where the guys wrote songs like 'She Loves You'.
HOW The 2.5-hour tours depart four times daily Wednesday through Sunday, March to November. See www.nationaltrust.org.uk.

CASA AZUL
Mexico City, Mexico
Beloved Mexican artist Frida Kahlo was born in, and lived and died in, the Casa Azul (Blue House), now a museum in Mexico City. Her soul permeates the rooms, which are littered with mementos and personal belongings.
HOW Arrive early at Casa Azul (www.museofridakahlo.org.mx) to avoid the crowds.

© ANTON IVANOV | SHUTTERSTOCK. © BONDROCKETIMAGES | SHUTTERSTOCK

A beer with Cesária

"Cesária Évora was the reason we went to Cape Verde. Her songs sold us on the idea of the romantic land. Not that we expected to see her. She was an international recording star, after all, who travelled the world crooning about her country.

So imagine our surprise when we were walking down the street in the town of Mindelo, and we saw Cesária Évora sitting on a porch. Her porch, apparently. We waved. Imagine, then, our greater surprise when she invited us in and offered us a beer.

We couldn't say much to each other directly, since we don't speak Criolo or French and Cesária doesn't speak English. Everything had to be translated by Peter, Cesária's 87-year-old uncle, a dapper gentleman in white linen trousers, as sharp-dressed as they come (except for his unzipped fly).

Cesária asked why we had come to Cape Verde and where else we had visited in the country. As Peter relayed our answers, she listened graciously, puffing on Marlboro Reds held between fingers heavy with gold rings.

At the end of our visit, Cesária let us take photos (after telling Peter to zip up first) and invited us to come back. She was a class act all the way."

Karla Zimmerman

BUILD UP TO IT
Practise daily gratitude, with paints:
Keep a sketch journal, p48
Travel by example:
Have adventures with children, p68

FLY

BUILD UP TO IT

Embrace a buoyant
blue world:
Spend time on water, p14
Find a head
for heights:
Do look down, p134

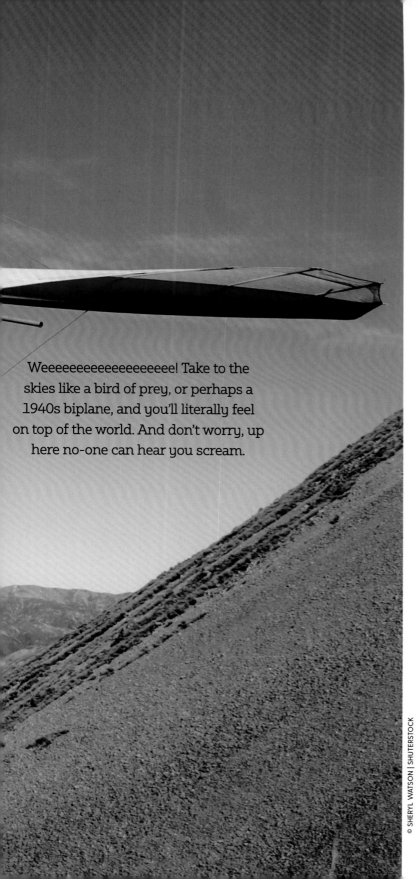

Weeeeeeeeeeeeeeeeeee! Take to the skies like a bird of prey, or perhaps a 1940s biplane, and you'll literally feel on top of the world. And don't worry, up here no-one can hear you scream.

WINGWALKING

Cirencester, UK

Wingwalking – the outrageous act of standing atop a biplane in flight – is a literal blast. You'll feel the onrushing wind as you soar high, skim low, bank and barrel; you'll plunge so steeply that you're suspended weightless.

HOW Fly above the Cotswold hills with Aerosuperbatics (www.aerosuper batics.com).

HANG-GLIDING

Kitty Hawk, North Carolina, USA

Learn to hang-glide in the home of flying: it was from Kitty Hawk's sandy beach in 1903 that the Wright brothers made the first powered flight. Kitty Hawk Kites gliding school will teach you to launch off the dunes just like they did.

HOW Two-day beginner courses include a tandem flight. www.kittyhawk.com.

PARAHAWKING

Algodonales, Spain

Paragliders need air thermals, and at this unique spot black vultures and Harris's hawks guide paragliders towards the rising air, soaring alongside them and even perching on their gloved wrists.

HOW Skywings Falconry (www. parahawking.com) offers tandem parahawking from October to May. It is 56 miles (90km) from Seville.

ZIP-LINING

Toro Verde, Puerto Rico

Strung over lush Caribbean rainforest, the Monster is one of the world's highest, longest and fastest zip-lines. It takes about 150 seconds to career for 1.57 miles (2.5km) at 1250ft (380m) above the valley, reaching speeds of up to 95mph (153km/h).

HOW Toro Verde (www.toroverdepr. com) offers zip-lining on the Monster, 40 miles (65km) southwest of San Juan.

Left: Hang-gliding in New Zealand

When the modern world's constant clamour gets your ears ringing, escape on a silent retreat and take your meditation to the next level.

EXPERIENCE A

*D*edicating time to purposeful silence is practised in many cultures and religions, from Trappist communities to Zen Buddhists. Often observed as a way to gain closer communion with a god, prolonged silence, especially in the company of others, is also understood to deepen self-awareness and gain clarity. The pre-Buddhist technique of Vipassana, which means 'to see things as they really are', involves silence, stillness, and observing your breath as a way to eradicate mental impurities. Practitioners learn the technique at 10-day silent meditation retreats, held at some 160 centres worldwide.

The growth of the wellness industry has seen silent yoga retreats and mindfulness meditation centres pop up the world over – from Los Angeles to Bali and South Africa to Devon in the UK – meaning finding space for quiet has never been easier.

The hard part comes in seeing it through. Silence isn't always golden. This is the lesson many first-time silent retreaters speak of. Silence is hard, and the first day will usually be spent resisting the urge to run screaming for the door or crumble in a fit of the giggles. And then what? Then comes the deafening chatter of your internal monologue. Being in a room of silent people can feel deeply strange. Yet devotees will argue that if you can make it through, a week of silence can be transformational, with rewards including better sleep, reduced anxiety and stress levels, improved focus and a better memory.

WEEK OF SILENCE

BUILD UP TO IT

Switch off and be present:
Leave your smartphone at home, p80
Learn from a spiritual leader:
Meditate with masters, p216

KYOTO KUKUSAI ZENDO
Japan
Get Zen – for real – at this traditional Japanese Zen centre. Guests come from all over the world to practise *zazen* – the art of seated meditation – in the *zendo's* incense-fragrant wooden buildings, set in the hills outside Kyoto. Silent meditation is only broken by ritual chanting, though chatting is allowed in the evenings.
HOW The centre is located in Kameoka, 30 minutes by train from Kyoto. Information on how to apply is given on the website, www.tekishin.org.

DHAMMA GIRI
India
Dhamma Giri, in the Indian state of Maharashtra, is one of the world's largest Vipassana meditation centres. Meditate beneath cyan skies in a golden pagoda with more than 400 separate meditation cells, and share vegetarian meals with everyone from cleaners to CEOs. Adherents claim benefits that range from hallucinations to full-body orgasms.
HOW Dhamma Giri is in Igatpuri, a three-hour drive from Mumbai. Apply for a 10-day residential course through the website, www.dhamma.org.

HRIDAYA YOGA
Mexico
Practise your downward-facing dog and tree pose in blessed quiet at Hridaya, a yoga and silent retreat centre on Oaxaca's glorious Mazunte beach. Get your toes wet with a three-day retreat, hone your fortitude with a 10-day session, or go full-on hermit with an invitation-only 49-day solo experience.
HOW Mazunte is an hour's drive from Puerto Escondido, or six from Oaxaca City. Book through the website at least a month in advance. www.hridaya-yoga.com.

Left: A Japanese lily pond offers a calming backdrop

MASTER A FO

BUILD UP TO IT
Take a doorstep challenge:
**Be a tourist in your
own country, p34**
Engage with the unknown:
**Immerse yourself in
local culture, p88**

REIGN TONGUE

Break down barriers and connect with a culture like never before, because a little language-learning can go a very long way.

Maybe you studied Spanish in school for years, but never managed to learn the difference between 'ser' and 'estar'. What's more, you probably didn't really care. It wasn't until you spent a semester in a small, dusty desert town – let's say, San Martin in Argentina – at the age of 17 that the language took root in your brain and began to blossom. Because you discovered that learning a language isn't just about conjugating verbs. It's also about drinking maté tea with new friends and dancing at the discoteca, and chatting with the bakery ladies as you buy your dulce de leche pastries.

There are so many reasons to learn a new language. It unlocks new cultures and new experiences. It keeps your brain youthful and active. It even opens up professional opportunities: those who speak Mandarin as a second language, for example, find themselves in high demand. Snoozed your way through high-school Spanish class? Make up for it with a Spanish immersion course in the handsome colonial city of Oaxaca in southern Mexico. The Instituto Cultural Oaxaca (ICO) offers Spanish classes, homestays with local families, and cultural courses on topics from cooking to salsa dancing to piñata making.

CANTONESE
Sure, Mandarin may be the dominant Chinese dialect, but Cantonese has better curse words! Plus, you can watch Bruce Lee movies without the subtitles. **HOW** The Hong Kong Language School, in neon-lit Wan Chai, holds classes ranging in duration from two to 16 weeks; www.hkls. com.hk.

DANISH
Want to read Kierkegaard in the original? Study Danish at Studieskolen in the heart of Copenhagen. It's got the royal seal of approval – the Crown Princess of Denmark, an Aussie who met her husband in a Sydney pub, learned Danish here. **HOW** New courses start every six weeks. See www.studieskolen.dk.

HINDI
A great place to learn the world's fourth most widely spoken language is Language Must in Delhi, where native speakers guide you through basic conversation skills. After just a few lessons, you'll be ordering your *jalebis* (syrup-soaked fried pastries) and *chole bhature* (spicy fried chickpeas with bread) like a local. **HOW** Group and one-to-one classes are available; www.knowledge-must.com.

SWAHILI
Commonly used throughout southeast Africa, Swahili is considered relatively straightforward to learn. Find out whether or not this is true with a Swahili course at the Language Centre in Nairobi. Spend the morning studying, and take the afternoon to roam the city putting your skills to the test. **HOW** Study options range from private hourly lessons to daily group courses; www.language-cntr.com.

Left: Take your new skills to the streets of Copenhagen, Denmark

DISAPPEAR

Going off-grid for a few weeks in a remote wilderness can be
a mind-altering, life-affirming experience. With only herself
for company and the changing elements as constants, **Kerry
Christiani** finds solace in the Outer Hebrides, where life is
more elemental and intuitive.

The Outer Hebrides inhabited my imagination long before I ever set foot on these wee specks of treeless islands in Scotland's far northwest. I was drawn to their edge-of-the-world quality, Gaelic spirit and the kind of solitude you can only get in the back of beyond, where there's no wi-fi and you have to walk a stiff mile for a patchy mobile signal. Needing to disconnect completely, and with five weeks on my hands, this sounded like utter bliss. My friends were aghast: five weeks alone? Wouldn't I go stir crazy? Wouldn't I get bored? What on earth would I do? Nothing. I would do absolutely nothing. That was the point.

And so it was that I booked a tiny hideaway croft: a sagging, 200-year-old stone cottage with a thatched roof, low beams and a wood-burning stove. It nuzzled in the crook of a small bay on the island of Berneray, looking out across the glimmering Sound of Harris. Population: 136. The light was swiftly fading on the autumn day I arrived, but I could tell it was perfect, and lovelier in reality than I had ever dreamt it would be. Below were rocks slick with seaweed, beyond were bare hills, the sea and the sky. Aside from the wing beat of barnacle geese flying in a V-formation overhead, it was totally silent.

Over the coming weeks, with no plans or responsibilities, my days fell into a natural pattern in tune with the rhythms of the island. In the morning, I'd sit on a rickety bench outside, mug of tea in hand, to watch the tide going out, and would occasionally be joined by the odd seal, announcing its arrival with a friendly snort and a glossy head emerging above a quicksilver sea. Then I'd put on boots to wade down to the shoreline at low tide to observe the rock pools, quivering with miniature life: vivid green algae and jewel-like sea anemones, encrusted with periwinkles, whelks, limpets and barnacles.

Beyond the rock pools there was a magnificent and empty coastline to explore. Up I would climb to the sheep-grazed rise of Ben Leva, past dry-stone walls and down the other side to West Beach, a surreal 3-mile (5km) expanse of white sand and sea ranging from turquoise to deep sapphire, fringed by dunes cloaked in machair – a lush sea grass formed by crushed shells blown ashore. When walking here I'd lose track of all time, the hours marked solely by the changing light and tides. I'd return at sunset, with peachy-gold and bruised purple-pink skies rendering the island into silhouette.

There was a beautiful simplicity to my life here. At low tide I might forage for cockles or mussels, heading out with rake and bucket. When the weather turned, I'd huddle inside with the radio on and peat fire burning, watching the waves being whipped into a frenzy and the rain come down in lashing curtains, draining the landscape of all colour. An hour or two later, it might have blown over and the island would be drenched in brilliant sunshine, a rainbow arcing over the grasslands.

One evening, close to the end of my time on Berneray, I stepped out to observe the stars and there were the Northern Lights, flickering like green and pink strobes in the night sky. I walked elated into the dark night, completely alone and watching one of the greatest shows on earth. It was the most moving of goodbyes. Closing the croft gate for the last time on a dreich November morning, I felt my heart sink just thinking about going back into a world so far removed from this one. Because, ultimately, disappearing to wild places doesn't take us away from reality, it brings us closer to it.

HOW The quickest way to reach Berneray is by flying from Glasgow to Benbecula; from there it's an 1.5-hour drive. For holiday rentals, see www.isleofberneray.com and www.homeaway.co.uk. Winters can be bleak – try spring to autumn (May to October).

"Over the coming weeks, with no plans or responsibilities, my days fell into a natural pattern in tune with the rhythms of the island."

FAROE ISLANDS

Technically part of Denmark, these magical isles midway between Norway and Iceland have a remote, edge-of-the-world quality. Fickle weather is to be expected, but holing up in a turf-roofed cottage is a brilliant way to slip off the radar for a while. The Faroes have an almost primordial beauty, with their wave-lashed sea cliffs (look out for puffins), layer-cake mountains and shimmering fjords.

HOW Atlantic Airways has flights between (among other destinations) Edinburgh and Vágar. You'll need your own set of wheels to explore (car or bicycle). Holiday rentals are listed at www.homeaway.co.uk, www.airbnb.co.uk and www.holidaylettings.co.uk.

RAPA NUI (EASTER ISLAND)

It doesn't get more middle-of-nowhere than Rapa Nui, plonked in the mid-Pacific, 2300 miles (3700km) west of Chile. This island is an all-round mystery, with Polynesian roots, monolithic maoi statues wearing topknots, extinct volcanoes, crater lakes and white-sand beaches. For solitude, avoid peak season (January to March) and explore with a backpack and camping gear.

HOW LATAM flies between Santiago de Chile and Easter Island. Wild camping is forbidden (or at least frowned upon) as much of the island is national park. Official campsites are best; try Mihinoa: www.camping-mihinoa.com.

CHRISTMAS ISLAND

If you'd rather forget the whole turkey-and-present-giving thing, where better to go off-grid during the festive season than Christmas Island? Some 220 miles (350km) south of Java, this Indian Ocean island has a rainforest-swathed national park, high sea cliffs, coastal trails leading to palm-rimmed, sugar-white beaches, and wildlife from the miniature (migratory red crabs) to the massive (whale sharks).

HOW Fly here from Perth, Australia, or Jakarta, Indonesia. Rent a lodge by the sea for near-total solitude; www.christmas.net.au lists some options.

FACE
YOUR FEARS

Don't be driven by comfort or fear. It's just that simple. We all get scared and crave comfort, but if we set up our lives to avoid what terrifies us, or to always choose to be nurtured rather than challenged, how will we grow?

*N*othing challenges us to the core like coming nose to nose with some terrifying proposition. Fear gets the stomach churning and mind whirring as panic builds. When that happens most of us want nothing more than to retreat. But each time we back down or avoid something, rather than face what scares us, our fears only grow stronger, and our mind gets a little bit softer.

That doesn't mean if you're scared of and uncomfortable in the ocean, or are inexperienced in the mountains, you should force yourself out past the waves, or up a steep slope without preparation or support. Because fear can produce panic, and panic can be deadly in some environments.

But for those of us with irrational but very real fears, facing that fear up front in a safe environment and with expert help can be the most empowering experience you're ever likely to have.

So much that it may be enough to live by the following rule: if you come across something you don't want to do because it scares you, go do that thing. When it comes to fears, often the only way out is through. In time, what once terrified you might become something you actually enjoy doing.

SWIMMING WITH SHARKS
Guadalupe Island, Mexico
Humans are hardwired to be terrified of
sharks. Never mind the fact that you have
a far better chance of getting struck by
lightning than getting nibbled by a man-
eater. Dive operators all over the world
can get you in the blue with harmless reef
sharks, but why not cage dive with Great
Whites off Guadalupe Island in Mexico.
The biggest and best sharks gather in
September and October.
HOW Horizon Charters (www.
horizoncharters.com) organises five-day
trips departing from San Diego.

GLIDE THROUGH THE SKY
Ölüdeniz, Turkey
Do you fear heights? Can't stand flying?
Go tandem with a pro paraglider and let
sheer exhilaration overwhelm your duelling
terrors. Clench your eyes shut at first if you
must, then open one at a time and gaze to
the hills and the sea.
HOW Sky Sports (skysports-turkey.com)
in Ölüdeniz, Turkey, launches from coastal
mountains over the Mediterranean. Fly into
Dalaman Airport, a 1.5-hour drive north.

DANCE YOURSELF FREE
Havana, Cuba
Although seldom placed high on the list of
phobias, every wedding reveals its share
of wallflowers afraid to expose their
groove. There's only one antidote: learn
some moves! Havana's all-ages dance club
culture and abundant dance studios
make it the perfect place to start.
HOW Salsabor a Cuba (www.
salsaborcuba.com) offers
personalised instruction
in their Havana dance
studio seven days a week.
Instructors will send you
spinning wherever music
plays.

Left: Cage divers meeting a female Great
White off Guadalupe Island, Mexico

David Goggins on callousing the mind

David Goggins, author of *Can't
Hurt Me*, transformed himself
from an obese, insecure young
man, afraid of heights and
deep water, into a US Navy
SEAL, champion ultramarathon
runner and an expert high-
altitude skydiver.

'I did it by callousing my
mind, callousing over my fears,'
Goggins says. 'The first step on
the journey toward a calloused
mind is stepping outside your
comfort zone on a regular basis.
Doing things that make you
uncomfortable will make you
strong. The more often you get
uncomfortable, the stronger
you'll become, and soon you'll
develop a more productive, can-
do dialogue with yourself when
your deepest fears have you in
the crosshairs.'

BUILD UP TO IT
Get your blood pumping:
Feel the rush, p102
Listen out for the call of
the wild:
**Spend a night in the
jungle, p128**

MAKE AN EPIC OVERLAND JOURNEY

No flying allowed! Get from point A to a very distant point B the old-fashioned way: by car, train, bus, foot or yak-drawn cart, and learn why it really is all about the journey.

*U*s humans, we could easily just stay home. But we don't, do we? There seems to be something in our DNA that draws us to travel long distances. From our migration out of Africa nearly 200,000 years ago, to ancient peddlers travelling the Silk Road, to 19th-century Westward Expansion along the Oregon Trail, we just can't stop moving.

These days, we've got jets to get us to wherever we want to go. Marco Polo's trip to China took some 20 years. Now we can fly from Italy to Beijing in less than half a day. But does that diminish the value of epic overland travel? Not at all. If anything, it makes land-based trips feel more special. Anyone can book a plane ticket, but it takes planning, thought, time and a love of serendipity to figure out how to get across an entire country or continent using local transportation.

Nobody knows this like Lonely Planet's founders, Maureen and Tony Wheeler, who launched the company after a (mostly) overland odyssey from London to Sydney on the 'hippie trail' through Europe, the Middle East and Central Asia. It inspired them to write their first title, *Across Asia on the Cheap*. The rest is history.

Right: Samarkand, Uzbekistan

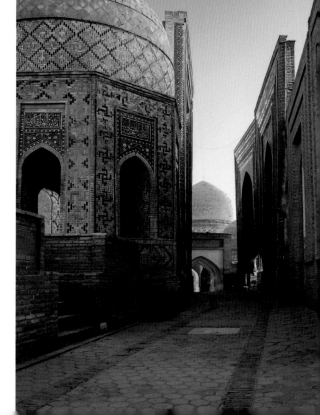

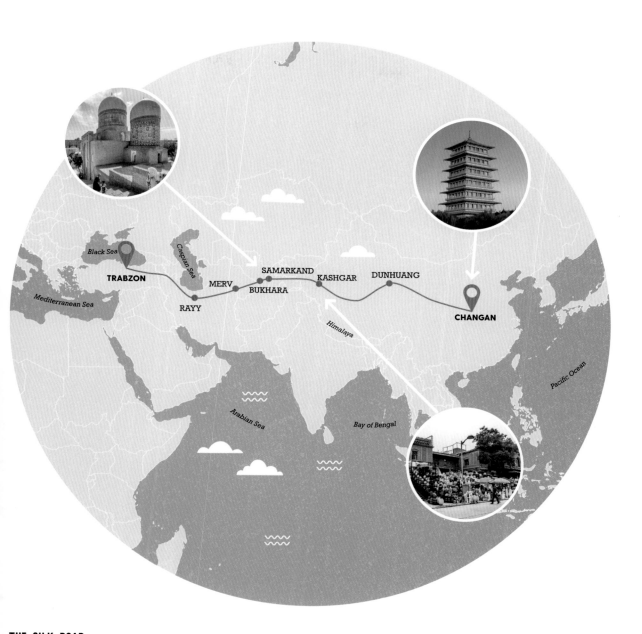

THE SILK ROAD

China to Turkey

Take an exotic journey across Central Asia, threading through former Soviet Republics and past Islamic architectural marvels, on roads once traversed by caravans carrying silk worth its weight in gold. Modern travellers will likely make the same route decisions as early traders, based on cost, ease of transport and the time of year. Highlights of the route include the Kashgar's Sunday market, medressa-hopping through Samarkand and Bukhara, and touring the scattered ruins of Merv.

HOW Most travel is by train, or hire car if you want to explore areas away from major cities. The biggest headache of the trip is visas: Kyrgyzstan and Kazakhstan are visa-free, Turkey and Tajikistan have an easy online process, Uzbekistan and China are fairly easy and Turkmenistan and Iran are tricky. **Duration: three to four months; distance 5000 miles (8000km).**

THE PAN-AMERICAN HIGHWAY

Canada to Argentina

From the tundra of Prudhoe Bay, Alaska, down to the equally icy tip of Argentina's Tierra del Fuego, this 30,000-mile (48,300km) driving route defines 'epic'. The US portion has multiple branches, so simply choose your own adventure. Then skitter through the capitals of Central America, ditching your car or bus in Panama to cross the roadless Darien Gap by boat or plane and picking up another in Colombia for the ride down South America's West Coast. Along the way you'll rumble through the remote Arctic north, glimpse epic Mayan ruins and hear the calls of howler monkeys, roll through grasslands and over steep Andean passes before reaching the exquisite desolation of Tierra del Fuego.

HOW This route is best done with your own car or with local buses. See www.go-panamerican.com.

Duration: at least six months; distance: 19,000 miles (30,000km).

CAIRO TO CAPE TOWN

Egypt to South Africa

Got four months to spare? Hit the most glorious bits of the African continent with an up-to-down road trip (or down-to-up, if you prefer). You'll take in the pyramids, dive the Red Sea, hike the cloudy foothills of Mt Kilimanjaro, hear the thunder of Victoria Falls and spot basking lion prides in South African wildlife reserves. Depending on your route, you'll hit about a dozen countries. Must-brings: a 4WD and patience.

HOW Travel in your own car or by guided tour with an operator like Dragoman Overland (dragoman.com) **Duration: six months; distance: 6284 miles (10,114km).**

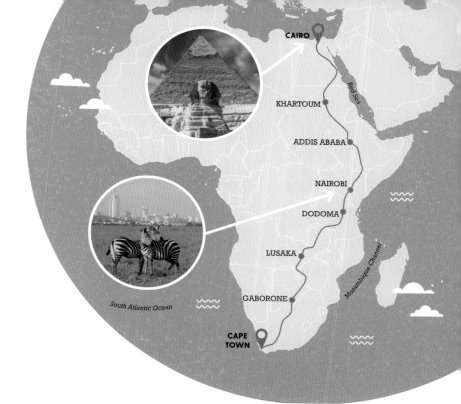

CAIRO

KHARTOUM

Red Sea

ADDIS ABABA

NAIROBI

DODOMA

LUSAKA

Mozambique Channel

GABORONE

South Atlantic Ocean

CAPE TOWN

DARWIN

Coral Sea

CAIRNS

PORT HEDLAND

BRISBANE

PERTH

SYDNEY

ESPERANCE

ADELAIDE

Ocean

MELBOURNE

NEXT 96 km

HIGHWAY 1

Australia

See just how big Australia really is by driving a giant circle around it. From Melbourne, go counter-clockwise to Sydney, then up the mango-scented Queensland coast to Cairns for a last-chance dive on the Great Barrier Reef. Then west to Darwin and then follow the coast all the way down to Perth before crossing the treeless Nullarbor into the welcome green of South Australia. If you don't have months to dedicate, just do a portion at your leisure.

HOW In your own vehicle, sleeping at campsites. If you're only doing a small portion, you could rent a camper van (www.britz.com) **Duration: one year; distance 9000 miles (14,500km).**

LIVE OFF THE LAND

For most travellers, the local food and natural characteristics of a place are its magic ingredients. What better way to appreciate and sustain these treasures than by getting up close and personal with nature's bounty down on the farm?

The idea of living off the land can stir notions of early pioneers relying on their wits as they cart across vast wildernesses, or perhaps modern-day freegans come to mind, surviving off that which others have thrown away. That's not quite what's intended here, although it's certainly related. Instead, this is about the unique natural qualities of a place that should be appreciated and preserved, both by the locals who live there and by visitors.

Another word for it is 'permaculture', a portmanteau term combining 'permanent', 'culture' and 'agriculture'. It describes a broad set of practices supporting the value of working with nature, not against it. In fact, permaculture is a way of life to anyone committed to it, touching on more than just food and organic agriculture. From using cut grass and fallen leaves as garden mulch to sourcing energy sustainably, it's an ethical approach to living that benefits the planet and its people.

For travellers, popular variations on the permaculture theme include agritourism, which revolves around farm stays and other agricultural adventures, and WWOOF, a worldwide movement linking volunteers with organic agriculturalists. They're land and life experiences that every earth-conscious nomad should try.

ZAYTUNA FARM

Australia

Zaytuna Farm is the home of the Permaculture Research Institute, a not-for-profit organisation specialising in permaculture education, training and consulting. Zaytuna also hosts events and courses for the curious, and volunteer and apprenticeship programmes for skill building. **HOW** Zaytuna's 66-acre (27ha) farm in northern New South Wales offers seasonal tours and courses in permaculture. See www.zaytunafarm.com and www. permaculturenews.org.

AGRITOURISM

Italy

Although agritourism is practised worldwide, it's closely associated with Italy, where it's widespread, regulated by law and usually high quality. Here, agriturismo is largely B&B-type overnights on farms where visitors get to know the people who live off the land and the food they produce. **HOW** Numerous websites, including www. agriturismo.com, catalogue agritourism in every region in Italy. Tuscany is famous for vineyard stays.

WWOOFING

Worldwide

Since 1971, WWOOF (Worldwide Opportunities on Organic Farms) has been arranging for people 'passionate about healthy food, healthy living and a healthy planet' to support agricultural efforts on sites in more than 60 countries. **HOW** It's an exchange of muscle and goodwill for room and board during a fixed period of time negotiated with a host. Get started at www.wwoof.net

© JAN MALKOVSKY | ALAMY STOCK PHOTO

Left: Vineyards at Greve in Chianti., Italy

Permaculture priorities in travel

Jonathon Engels, a well-established permaculture writer, believes incorporating permaculture into travel plans should be important to every traveller. 'An exemplary permaculture design has explanations for the minutiae, from housing that fits with the environment to crop plants that suit the climate,' he says. 'A lot can be learned about a place by visiting a permaculture site. Permaculture aims to keep the planet and its population hospitable. Travellers can begin as simply as buying locally, supporting ethical businesses, and visiting national parks and conservation projects. Basically, if we think of what is environmentally and people-friendly and go those routes, our travel can have a positive impact, as opposed to a potentially destructive one.'

BUILD UP TO IT

Get outdoors:

Sleep under the stars, p42

Consume thoughtfully:

Go meat-free on the road, p236

FIND YOURSELF

ASHRAM LIFE
Munger, Bihar, India
Yes, it's yoga. But at
the Bihar School of
Yoga, the focus is
karma yoga – the
path of selfless action.
That means hours of
service – think kitchen
work and gardening –
alongside your asanas.
A simple diet, 4am
starts, nightly mantra
chanting and silence
from 6pm to 6am will
take you far from your
everyday reality.
HOW Stay at the
Bihar School (www.
biharyoga.net) for one
week or four months.

WALK THE LAND
Uluru, Australia
For Australia's native
peoples, the spirits
of ancestral beings
reside in trees, rocks,
caves, boulders and
waterholes. Songs
describing the location
of the landmarks and
telling ancestor stories
can be followed for
hundreds of kilometres.
At once spiritual belief
and navigation system,
the songlines uncover
a profound connection
to the land.
HOW Visit Uluru's
Cultural Centre (www.
parksaustralia.gov.
au/uluru) to learn the
local ancestor stories.

EXTREME PILGRIMAGE
Mt Kailash, Tibet
Although Mt Kailash
is one of the world's
holiest places, sacred
to Hindus, Buddhists
and Jains, it welcomes
only a few thousand
pilgrims each year.
Its remote location
requires weeks of
difficult travel, and
around it are 32 miles
(52km) of path to
circumambulate – for
some pilgrims, in one
day; for others, with full
body prostrations the
whole way.
HOW Book at Tibet
Travel (www.tibettravel.
org); 15-day tours leave
from Lhasa.

SING YOUR HEART OUT
Seoul, Korea
Most people can't sing.
In Korea, that's the
whole point. Uninhibited
group singing is integral
to Korean culture –
your average night
out means going to a
noraebang (karaoke
room) and belting
out tunes. The more
unbridled you get, the
better. You're among
friends here.
HOW Go to a nightlife
area like Seoul's
Hongdae to choose
from countless
noraebang venues.

SACRED SHRINES
Kumano Kodō, Kii Peninsula, Japan
In ancient times,
Kumano, on the
mountainous Kii
Peninsula, was a sacred
forest where emperors
and samurai went to
worship nature. Shinto
temples and shrines
appeared, creating
the Kumano Kodō
pilgrimage trail. There's
a wealth of hiking
options, from hours to
days, easy to strenuous,
to enrich the soul.
HOW DIY the Kumano
Kodō, or go with a
local outfit such as
Kumano Travel (www.
kumano-travel.com).

BUILD UP TO IT

Relish the disconcerting:
**See a city in a
jet-lagged haze, p36**
Go to new depths:
**Descend into the abyss,
p206**

We travel to discover the world outside, but it can be the catalyst for an inner journey, too. Whether it's finding balance, experiencing revelation or overcoming limitations, there are many paths to personal growth. Choose a road and walk it with purpose.

SEEK VISION
USA
In Native American tradition, the vision quest is a rite-of-passage ritual undertaken by a young man. He ventures into the wilderness alone to fast and pray, trying to attain a vision of his future guardian spirit who will give him clarity on his life's purpose. A spirit we could all do with at some time in our lives.
HOW Learn how to do your own vision quest at www.native-americans-online.com.

SAY A PRAYER
Jerusalem
Jerusalem's Western Wall is a place of pilgrimage for the world's Jews. It's believed that prayers placed here have a higher chance of being answered – every crack between the ancient stones is filled with tiny folded paper prayers. Place yours among them and put your faith in a higher power.
HOW Dress appropriately. Women must cover their legs and shoulders, men their head.

LEARN TAI CHI
Yángshuò, Guangxi Province, China
Tai chi is many things: philosophy, martial art, meditation, gentle early morning exercise practised by retirees in parks around the world. But at its core is the cultivation of the life energy within us to bring total harmony of the mind and body.
HOW Train for anything from a week to a year at the Yangshuo Traditional Tai Chi School (www. yangshuotaichi.com) in Yángshuò, in the southern province of Guangxi.

LOSE YOUR INHIBITIONS
Helsinki, Finland
Sauna is a cornerstone of Finnish culture. Everyone is equal and nothing can be hidden when naked in a small steamy room heated to almost 100°C (212°F). Optional: a little self-flagellation with birch twigs. Recommended: forgetting your hang-ups and letting it all hang out.
HOW Enjoy the real deal at Sauna Hermanni (www. saunahermanni. fi), a public sauna established in Helsinki in 1953.

BE HAPPY
Copenhagen, Denmark
Sometimes, in our self-improvement frenzy, we can forget that it really has but one aim: to be happy. The Danes have found a short-cut. Ditch the punishing yoga regime, snuggle up and take simple pleasure in the present moment. It's hygge, and it can often be found in a cosy cafe.
HOW The Living Room, Larsbjørnsstræde 17, Copenhagen, is hygge heaven – couches, blankets, cakes and an open fire.

Above: Pilgrim on the Kumano Kodo, Japan

Consumer living will only get you so far. The rewards of a life well-lived cannot be bought. They don't come without pain or sacrifice, and there are few thrills on this sweet earth that compare with taking on an athletic challenge complicated by the natural world and nailing it.

The title here is meant only to capture your attention – it can't actually be done. Everybody knows Mother Nature is the most powerful physical force there is, and if you venture into her wilds hell bent on surfing head-high waves, climbing a cliff face or summiting one of the world's highest peaks, you do so with her welcome and at her whim. There will be risk involved and a hefty toll to pay.

That toll is preparation. Climbing frozen waterfalls, or swimming down to 100ft (30m) and back on a single breath, takes a commitment to fitness. All grand adventures demand small daily steps on a path leading towards game day. Without the small victories, you won't be as physically or mentally fit when the time comes, and it will be harder to push past fear and your perceived physical limitations to conquer your goal. In fact, the daily work is the whole point of this challenge.

We are not here to encourage you to accomplish some audacious first, so you can humblebrag on Instagram. The reason to plan, scheme and prepare for an ultimate adventure is because it will challenge you to live more intensely, and leave the sedentary life behind. Remember, your initial goal needn't be superhuman. The point here is to stretch your own limits and get out into the wild as often as you possibly can. Your reward just might be a whole new lifestyle.

CONQUER NATURE

Build up to it

Don't let the weather get in the way of a good time:
Embrace the off season, p12
Respect where it's due:
See the magnitude of the earth's power, p100

ROCK CLIMBING

Rock climbing combines a host of fears and challenges into one terrifying yet empowering obstacle course of danger and fun. First, get to your local rock-climbing gym and take classes. That's how Alex Honnold (of *Free Solo* fame) got started. Once you've done that and put in hours at the gym, sign up with an outfitter and get outside.
HOW Yosemite Mountaineering School & Guide Service (www.travelyosemite.com) operates within the Yosemite National Park. It offers equipment, lessons for beginners through to advanced, and full-day guided climbs.

CLIMBING ONE OF THE SEVEN SUMMITS

Some big mountains demand rope skills. Mt Kilimanjaro, on the other hand, is a walk up, but the altitude can be daunting. Wherever you book your first summit attempt, prepare with a regiment of running or circuit training. Then hike a succession of tall peaks. Taking a training course before tackling your first peak can also be valuable.
HOW Adventure Consultants (www.adventureconsultants. com) in Wanaka, New Zealand, offers a Seven Summits Training Course that introduces you to the art of mountaineering and teaches technical skills.

FREE DIVING

Gili Trawangan, Indonesia
Instructors at Freedive Gili, on Indonesia's Gili Trawangan – accessible by speed ferry from Bali – will have you holding your breath for three minutes and help you swim down to 100ft (30m) on just one breath during the three-day intermediate training course. Before class begins, build up to regular long-distance swims in a pool. If you're fit, you'll have more fun and success under water.
HOW Courses run all year on Gili Trawangan, but crowds thin and visibility peaks in September and October. www.freedivegili.com.

Left: Free diving

DON'T STOP TRAVELLING

Rather than return from your travels with the holiday blues, take a chance on a career change that allows you to pursue your dream life on the road. Unchaining yourself from a stable job and home base requires grit and perseverance, but the rewards are a desk in the sunshine (metaphorical or literal) and a life on your own terms.

Whether the thought of trading in the day job for a life of exotic adventures scares or excites you, the reality is it may well be in reach. Recent studies show more and more people are freelancing and working remotely. And who can blame them? When you become your own boss, you also get to pick your own office, whether that's island hopping and answering your morning emails with a fresh coconut or setting up a longer-term base at a digital nomad hub. According to a 2018 report by MBO Partners, nearly five million Americans describe themselves as digital nomads, and as the numbers grow, so do the associated services. Over the past few years, co-working spaces, online job markets and community-focused travelling have boomed.

And if you're not sure you've got what it takes to be your own boss, there's a wealth of information and learning available online. Developing the skills needed to work as a virtual assistant, web designer, photographer or any other remote position has never been easier.

So what's stopping you? Travelling with an open ticket allows you to travel at a slower pace and gain a more in-depth understanding of the cultures around you. Also, by working in new environments and learning to adapt to new challenges, you'll be developing skills and expanding your CV while having a life-changing travel experience.

It's wise to make the leap with some savings in the bank, but keep in mind that even if your plans run out of steam, the worst that can happen is returning home having had an unforgettable adventure.

Left: Lisbon, Portugal

DIGITAL NOMADING

Southeast Asia
There are many digital nomad hubs with ready-made communities to integrate into in Southeast Asia. Chiang Mai in Thailand and Canggu in Bali, Indonesia, are two of the most popular, and these can be ideal first bases as you dip your toe into the life. **HOW Dojo in Canggu, Bali (www. dojobali.org) has plenty of networking events and an onsite pool; it's a great place to build your community alongside getting your work done.**

CO-WORKING

Europe
If Europe is calling, then there are a few cities that are embracing digital nomadism: both Lisbon in Portugal and Tallinn in Estonia are home to popular co-working spaces with prices more friendly on the wallet than other European capitals. **HOW In Estonia, you can apply for E-Residency, which may allow you to open an EU online company, should your business model require it.**

LANGUAGE TEACHING

Worldwide
Teaching your local language to foreign learners has long been a way for globe-hopping travellers to make an income on the road. Lessons can be taught online via video call, or in person, though longer-term contracts usually need some forward planning and visa arrangements. **HOW The TEFL website (www.tefl. com) is a great place to look for English-language teaching positions. Most require a qualification such as a TEFL certificate or CELTA, though requirements vary.**

From waiter to world traveller

"My original plan on leaving London had been to work in restaurants in Australia on a working holiday visa, but I quickly became addicted to travelling and I didn't want to let that lifestyle go. I needed to learn some skills that would allow me to continue.

When other people in the hostel I was staying at went out partying, I'd stay in to study videos online and learn how to take better photos, build websites and create a travel blog, all new skills and passions of mine.

My first job opportunity came with a tour company that took me across Cambodia and Vietnam to create content, and over the next few years I was able to keep travelling and supporting myself through the online skills I'd taught myself, such as photography, travel writing and designing websites.

There were times when my income would dwindle and I didn't know whether any more work would come up. But knowing the alternative was to go home and give up on this dream kept me determined to succeed. I'll be forever grateful that I took the leap of faith."
Daniel Clarke, photographer and blogger

BUILD UP TO IT
Find the exotic close by:
Be a tourist in your own country, p34
Express your love of the open road:
Write a travel blog, p158

GIVE A YEAR OF YOUR LIFE FOR OTHERS

In our unequal world, any desire to help others is laudable. Engaging in acts of generosity on trips is what gave rise to volunteer travel. These days, long-term volunteer commitments, lasting a year or more, are finding favour. Remember: slow-growing trees bear the best fruit!

Over many years, international volunteers have donated countless hours to disadvantaged communities worldwide. Recently, however, serious questions have been asked about the real benefits of such altruism, particularly when it comes in short bursts and from people unskilled in the services they're offering. Reactions to the concerns have focused on the significance of responsible volunteering. For example, new notions emphasise better balance between action and edification, stressing learning before service. As big-hearted travellers accept that volunteering is as much about what they gain as what

they give, they should take meaningful time to settle into a place before attempting to help it.

As a result, more people are committing to extended periods of volunteer time – a full year or longer. These timelines allow for immersion, introspection and skill building. What better way to learn a language; understand social, political and environmental opportunities from new perspectives; and help develop locally pertinent solutions to persistent problems?

So travellers should take great care to identify organisations that embrace this understanding of ethics and best practices in volunteering.

VOLUNTARY SERVICE OVERSEAS (VSO)

Africa, Asia and the Pacific
VSO is a leading international development organisation that places thousands of volunteers of all ages and backgrounds in communities striving to improve their futures through better access to healthcare, education, food and income.
HOW VSO oversees volunteer efforts in 24 countries across Africa, Asia and the Pacific. See www.vsointernational.org.

RALEIGH INTERNATIONAL (RI)

Central America, Asia and Africa
This UK charity helps young people make a real and lasting difference in the world through volunteer work. Its projects focus on areas such as supplying safe water, building stronger communities and protecting the world's resources.
HOW RI is active in Costa Rica, Malaysian Borneo, Nepal, Nicaragua and Tanzania. See www.raleighinternational.org.

Below (left and right): Helping on a conservation programme in British Columbia, Canada; feeding gibbons at the Gibbon Rehabilitation Centre, Thailand

GLOBAL CITIZEN YEAR (GCY),

Central and South America, Asia and Africa
This US-based organisation recruits and trains young people for a gap-year 'fellowship' between high school and college. A new generation of leaders develops valuable new skills by living with local families and working as apprentices.
HOW GCY's young leaders are active in Brazil, Ecuador, India and Senegal. See www.globalcitizenyear.org.

PEACE CORPS

Worldwide
The Peace Corps places volunteers at the community level and has a focus on integration and sustainable change. Placements are 27 months and could be in one of 60 countries.
HOW Any US citizen over 18 can apply. See www.peacecorps.gov.

© HERO IMAGES | GETTY IMAGES. © AUSTIN BUSH | LONELY PLANET

Win-win work exchanges

"I travelled for many months in South America (Chile, Peru and Bolivia) because I wanted a change and it was the right moment in my life. I organised my experiences through Workaway (www.workaway.info), which enables win-win work exchanges, a perfect forum for travellers to go deeper into a culture and for hosts to get the practical help they need. I chose this because some people believe that volunteering is altruistic, about helping others. The reality is that they're helping you, too. It's an equal exchange: you're giving something, but you're taking something away.

I knew I wanted physical, outdoor work and greater connection with nature. It was really good for me. Most of the time I chose organic or permaculture farms and normally stayed for two or three weeks. I looked up positions that gave me a good sense of a country.

My experience confirmed many things: I really love immersive, lifelong learning opportunities; I'm incredibly resilient (which is a good thing to know about yourself); I strongly believe in the kindness of strangers; and I can keep up with young people, so many of whom I came away with such huge respect for. There are amazing and brave young people out there, doing incredible things."
Jane Carpenter

BUILD UP TO IT
Give back to your own community:
Volunteer at home, p74
Focus your concern:
Help save an endangered species, p232

WITNESS A MIRACLE OF NATURE

Observing a bona fide miracle of Mother Nature's conjuring, as it unfolds right in front of you, is arguably the pinnacle of travel.

*N*ow that we're all devoted disciples of David Attenborough, it's tempting to think you don't need to move too far from a TV-facing sofa in order to get an eyeful of one spellbinding natural phenomenon or another. But positioning yourself in the perfect place at exactly the right time to experience a unique organic event with all your senses is something that will stay with you forever. Some such occurrences have long influenced tourism trends – the Northern Lights in the Arctic winter, or great dramatic migrations across the African Serengeti and Masai Mara plains, for example; or the springtime explosions of plum and cherry blossom in Japan, where festivals are held to celebrate the colour-splattered occasion, or salmon season on Kodiak Island in Alaska, when grizzly bears can be seen fishing in streams. Other natural occurrences and encounters are rendered extraordinary by the odds stacked against you experiencing them. The wild populations of some animal species are so desperately depleted that catching a glimpse of them in their natural habitat would truly be miraculous – Arabian leopards in Oman's wadis, or bears in the Carpathian mountain of Transylvania – but you can join expeditions to try and make that magic happen.

HIGHLY EVOLVED SWIFTS

Cook Islands

On the remote isle of Atiu, a species of swift – the kopeka – occupy the Anatakitaki Cave. Endemic to the island, the birds have evolved to use sonar navigation while flying in the darkness of the cave, before reverting to birdsong and visuals once in the light.
HOW Atiu (www.atiu.info) is a short flight from Rarotonga. Tours to Anatakitaki Cave are run via Atiu B&B (www.atiutours accommodation.com).

FALLS GOLD

Yosemite National Park, USA

For some 10 days in late February, the setting sun is at such an angle that it illuminates Horsetail Falls in Yosemite, seemingly making the waterfall a cascade of molten lava.
HOW Parking in the 'FireFall' viewing zone (from Yosemite Valley Lodge to El Capitan Crossover) is restricted, but you can walk in from other areas. See www. yosemitefirefall.com.

Below left and right: Horsetail Falls in Yosemite; a nudibranch slug

COLOURFUL NUDIS

Scotland, UK

So many marine experiences are about the titans of the deep – whales, dolphins and sharks – but nature is just as interesting in micro. Take the nudibranch, in essence a salt-water slug, but with an array of body designs that are kaleidoscopically colourful.
HOW The Scottish Nudibranch Festival (www.facebook.com/scottish.nudibranch. festival) takes place from May to June.

LIVE ERUPTIONS

White Island, New Zealand

In New Zealand, where the land often seems to be coming apart at the seams, you can watch the planet being reshaped on White Island, where an active volcano huffs and puffs away.
HOW Five- to six-hour boat tours to White Island (www.whiteisland. co.nz) leave from Whakatāne.

© GREGORY B CUVELIER | SHUTTERSTOCK, © JOE BELANGER | SHUTTERSTOCK

Shark date

"Typically, when someone yells 'shark!' that's when you want to get out of the water. But this wasn't a typical day. I was in Western Australia to experience something special, and the shark shout triggered a rush to find fins, don dive masks and leap into the deep as quickly as possible.

Every March, exactly seven to 10 days after the appearance of the full moon, the myriad kaleidoscopic corals that comprise Ningaloo Reef off the coast of WA erupt in a mass spawning event, clouding the Indian Ocean water with the raw ingredients for new life and pulling in gazillions of micro creatures such as krill and plankton. This attracts the usually elusive denizens of the depths, including bus-sized whale sharks and alien spaceship-esque manta rays. And that, in turn, lures experience-chasers like me, seeking a meeting with the very biggest fish in the sea.

Despite their enormity, and the yawning expanse of their mouths as they open wide to filter feed, whale sharks are entirely harmless to humans (so long as you avoid getting thwacked by their tail). Swimming along, eye-to-eye, with such an enigmatic animal is a humbling, truly miraculous experience."

April to June is the best time to visit Exmouth, Western Australia, to swim with whale sharks.
Patrick Kinsella

BUILD UP TO IT

See epic landscapes:
Be awestruck by nature, p122
Feel small:
Meet the planet's giants, p136

HELP A COMMU

Natural disasters seem to be hurtling forth two at a time. Earthquakes, wildfires, tsunamis and hurricanes dominate newsfeeds the world over, and thanks to social media we can see the apocalyptic aftermath in real time. The initial shock usually gives way to a nagging question: how can I help?

It isn't always as easy as booking a flight into the hot zone and putting in some manual labour, especially in the developing world. The last thing you want to do is take an opportunity for paid work away from someone who may have just lost everything they own. It's better to be clinical about where and how to pitch in. If you're a skilled medic, engineer, teacher, fundraiser or documentarian, there may be an avenue to apply your skills in a fluid disaster relief landscape.

There will be immediate healthcare needs, of course. If schools have been destroyed then teachers can still play a role in keeping children focused on learning and growing despite the chaos around them. Skilled engineers may find a way to insure better building

practices, and get water, waste and energy systems up and running. There will be financial needs too, not just in the immediate term, but for years to come. That's where the storytellers and fundraisers come in. Showing up with cameras rolling can be a powerful way to capture international attention, and if paired with online fundraising campaigns, can help funnel funds strategically.

It's vital to have developed inroads via existing aid organisations or through local contacts on the ground before turning up. If you arrive with supplies, funds and a reliable network on the frontlines you will find enough success, and make enough mistakes, to inspire you help out again when duty calls.

BUILD UP TO IT
Spend your money wisely:
**Support a social
enterprise, p110**
Spread the love:
Pay it forward, p126

NITY REBUILD

RED CROSS

The Red Cross is the gold standard when it comes to first response after a natural disaster, and 90% of its work gets done by volunteers. But to qualify as one of its skilled technicians or as an emergency medical professional, you first need to apply, and it helps to bring a raft of skills tailored towards disaster relief to the application process.
HOW Apply online at www.redcross.org and receive training in disaster response, which will allow you to work in disaster relief shelters when tragedy strikes. Deployments vary depending on your skill set and the disaster's scope.

DIRECT RELIEF

Active in all 50 US states and across 80 countries, Direct Relief's mission is to improve the lives of those impacted by poverty and emergency. It has sent search and rescue teams, medical professionals and relief supplies to communities affected by California wildfires, earthquakes in Haiti and hurricanes in Puerto Rico.
HOW Apply directly with staff to begin volunteering at the office in Santa Barbara, California (www.directrelief. org), or by holding fundraisers on your own. Most of Direct Relief's work is in getting cash and medical supplies to community organisations on the ground.

DOCTORS WITHOUT BORDERS

Médecins Sans Frontières operates clinics all over the world, shows up on the ground at the site of natural and man-made disasters (like war) all over the world, and regularly recruits medical and non-medical professionals to its missions.
HOW Visit www.doctors withoutborders.org to get involved.

ALL HANDS AND HEARTS SMART RESPONSE

One of the most well-respected disaster-relief organisations in the world was founded in response to the 2004 Boxing Day tsunami in Indonesia. It hits the ground days or weeks after disaster strikes to repair or rebuild homes and vital infrastructure.
HOW A volunteer commitment of two weeks gets you a free flight into the disaster zone. See www. allhandsandhearts.org.

Above: Khao Lak, Thailand, was rebuilt after the 2004 tsunami with help from volunteers

Pay a visit to a traditional healer, and you will not only gain important insights about a culture, you may also benefit from the effects of rituals passed down through the years and the centuries.

CLEANSE YOU

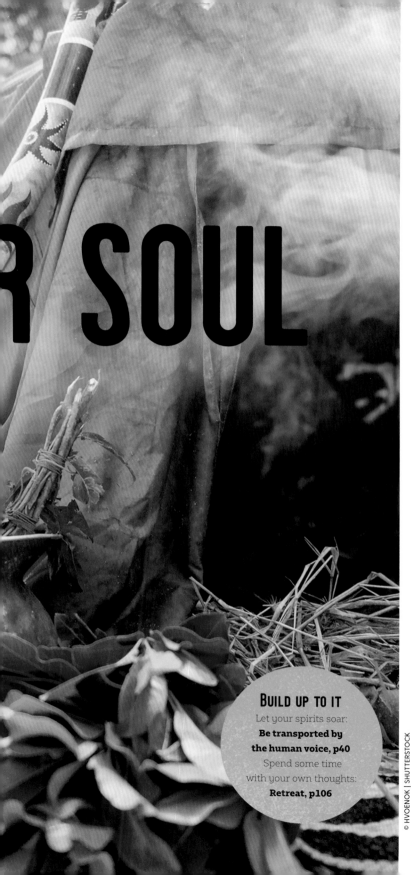

SOUL

MAYAN CLEANSING

Mexico

To the ancient Mayans, illness was not necessarily a physical ailment, but a spiritual one. They believed cleansing ceremonies could combat the supernatural. Today, many day spas and retreats in Mexico have fostered these rituals, with Mayan healers embracing ancient techniques. The *temazcal*, an indigenous Mesoamerican equivalent of a sauna or sweat lodge, is ritually used by the Maya to purify the body and spirit. **HOW** Yäan Wellness just outside Tulum holds 2.5-hour *temazcal* ceremonies in accordance with moon cycles. See www.yaanwellness.com.

BALINESE HEALING

Indonesia

Bali's traditional healers, known as *balian* (*dukun* on Lombok), play an important part in Bali's culture by treating physical and mental illness, removing spells and channelling information from the ancestors. Finding a *balian* is hard and etiquette should be strictly adhered to. **HOW** Bali Healers (www.balihealers. com/tours) is a good resource. It can select a *balian* for you to visit, while acting as liaison and translator.

HERBAL BLESSING

Hawaii

In ancient times, the *la'au lapa'au* (traditional healer) played an integral role in Hawaiian village life, often doing their work at specially dedicated temples. Today, traditional herbalists continue to practise, incorporating *lomilomi* (literally 'loving hands') massage into their treatments, which often include an *oli* (chant) sung for a blessing. **HOW** Skip the touristy beach resort spas and book an appointment with a local cultural practitioner such as Angeline's Mu'olaulani in Anahola, Kaua'i. www.angeineslomikauai.com.

BUILD UP TO IT

Let your spirits soar:
Be transported by the human voice, p40
Spend some time with your own thoughts:
Retreat, p106

© HVOENOK | SHUTTERSTOCK

Left: Mexican sauna hut

RESPECT THE CIRCLE OF LIFE

Getting that bit closer to death can be life-affirming.
Whether you're watching a lion go in for the kill, lighting a lantern to
show the spirits of your ancestors the way back home or celebrating
departed souls in macabre style with ghoulish masks and cloaks
of black, there's nothing like a near-death experience to help you
understand your own (im)mortality.

© SUSAN SCHMITZ | SHUTTERSTOCK

GREAT WILDEBEEST MIGRATION

Masai Mara, Kenya

Hook on to a Masai Mara safari during the migration season from July to October and you might sight a lion stealthily killing its prey: giraffes, zebras and, most dramatically of all, stampeding wildebeest.

HOW Rainbow Tours (www.rainbowtours.co.uk) offers an 11-day eco-safari package to the Masai Mara.

DÍA DE MUERTOS

Mexico

Mexicans remember the departed at these eerie festivities in November, with *calavera* (skull) and *calaca* (skeleton) costumes and decorations, processions, graveyard tours, and much mariachi music, dancing and merry-making.

HOW It's at its spiritual best in indigenous southern Mexico. All are welcome to join.

O-BON FESTIVAL

Japan

Lanterns float down rivers or take to the skies at this festival to welcome ancestral spirits back to the realm of the living in August. Graves are cleaned, offerings are made at household altars and temples, and folk dances are performed to the rhythmic beat of taiko drums.

HOW Arguably the best of all the O-Bon festivities is the Daimonji Gozan Okuribi Fire Festival on 16 August, with its blazing fires atop the mountains of Kyoto.

LA FIESTA DE SANTA MARTA DE RIBARTEME

As Neves, Spain

At this strange festival held on 29 July in As Neves, people who have had near-death experiences are paraded through the streets in coffins, with their mock-mourning family in tow, to give thanks for having been spared from death.

HOW As Neves is less than an hour's drive southeast of the city of Vigo.

BUILD UP TO IT

Learn from the past:
Retrace the steps of history, p30
Meet a killer:
Look into the eyes of a preadator, p202

BE YOURSELF

A long way from home, in the depths of an unfamiliar culture, where no one knows your last name or cares where you come from or what you do for work, an opportunity presents itself: to truly be yourself. It's an elusive goal but one that **Clover Stroud** managed to achieve in the remote unknown of Ossetia.

*B*y the time I had hit my late twenties, my first, eventful, but short-lived, marriage was over and I was a single mother to a toddler and baby. It was shocking, but also exhilarating, since the future lay entirely in my hands. I was working as a freelance journalist and wrote in the pockets of time I found when the children were napping, or late at night after they'd gone to bed. Life as a single mother was precarious, but also exciting, and as a travel journalist I was also lucky to have regular trips away on my own, to contemplate my life and where I wanted it to take me.

Just before I turned 30, I met a Russian acrobat working for a summer in England. He seduced me with urgent, romantic stories of his homeland in North Ossetia, deep in the Caucasus mountains, in wildest, most remote southern Russia. When his visa ended and he asked me to visit him in Russia, I knew I had to go.

As a journalist, I also jumped at the chance to write about the Caucasus mountains, since it's an almost mythical region, sometimes described as the Garden of Eden, a place that very few Western tourists visit. Two months later, with my children spending school holidays with their father, I hugged them tightly then flew to Moscow, meeting my acrobat at Domodedovo Airport in Moscow, and from there we boarded a train to Vladikavkaz.

As the train snaked across the Steppe, I felt something inside me shifting. I had the chance to step right outside my normal, maternal life, bound by school runs and work deadlines, and into a completely new world. It was exciting, and as the train stopped at increasingly remote stations south of Moscow, traders boarded, selling gold chains, pots of honey, finely spun cashmere scarves and bags of spiced pastries. A day later, at Rostov-on-Don, soldiers bound for Chechnya boarded. The travellers I met on the train displayed a kind of unquestioning sense of hospitality and generosity I'd never really experienced before. There were card games, late-night shots of vodka, coffees on freezing platforms. My Russian was very basic, but it was, unquestionably, the most exciting, eye-opening journey of my life.

I went back to Ossetia many times over the next few years, snatching a week here or there when I could leave the children and had sold another article. I always arrived by train, since the two-day journey gave me a chance to decompress from the pressures of home and somehow become myself again.

Ossetia revealed itself to me as a land of high romance, framed by the mighty, snow-peaked Caucasus and the boundless generosity of its inhabitants, who were passionate about their homeland. My acrobat boyfriend was an excellent host, determined to help me experience everything he loved about his home, including colourful markets, treks high into the mountains, ancient Pagan traditions and its famed mountain hospitality.

Ossetia gave me the chance to step outside convention, and the demands of normal life, and experience the deep joy of real travel. But it also did something else: it allowed me to grow up. It reminded me how huge and beautiful the world really is, after a few difficult years, and in doing so made me reach for a bolder, braver version of myself. More than anything, it made me realise that in risk taking, I could also grow as a human – and while the romance didn't last for ever, that lesson has been invaluable.

TRANS-SIBERIAN RAILWAY
Russia
As the 6152-mile (9900km) trip from Moscow to Vladivostok is the least-popular Trans-Siberian train journey, it is the one on which you're most likely to have a real adventure and feel furthest from home. Befriend the *provodnitsa* (female guard) who runs the train, and tip well. Take thermal underwear for disembarking, but lighter clothes too, as the trains are hot.
HOW Organise through www.realrussia.co.uk. You can opt for either a first-class *spalny vagon*, second-class *kupe* or third-class *platskartny*.

HUSKY SAFARI
Finland
Exploring Finland on the back of a sled pulled by huskies, and learning a few skills while you're at it, gives you plenty of time to clear your head as you gaze at the wild landscape around you. Stay in remote cabins as you travel through the extraordinary Pallas-Yllästunturi National Park while you seek out the majesty of the Northern Lights. The best time to travel is from early winter until late spring.
HOW Responsible Travel (www.responsibletravel. com) offers eight-day trips including flights, accommodation and meals.

HIKING
Peru and Bolivia
Hike from La Paz to the lowlands of the Yungas and the heart of ancient Bolivia. There's something extraordinary yet humbling about stepping in the footsteps of the ancient Inca while crossing a bright, green landscape of misty hillside and remote villages to take you right into the heart of the jungle. Go year-round.
HOW Journey Latin America (www.journey latinamerica.co.uk) can organise an 18-day trip from La Paz, Bolivia, to the Altiplano.

Daring to be yourself

Authenticity – being yourself – is correlated with many aspects of psychological well-being, including vitality, self-esteem and coping skills. But authenticity isn't easy. It requires making conscious, informed choices based on accurate self-knowledge, accepting your weaknesses as well as your strengths and daring to be open and honest with your relationships. The rewards are many. Stephen Cope, author of *Yoga and the Quest for the True Self* explains: 'Real fulfilment comes from authentically grappling with the possibility inside you, in a disciplined, concentrated, focused way.'

Index

A

abandoned places & ghost towns 16–19
adrenaline & adventure 68–71,
        102–103, 114–115, 152–153,
        218–219, 230–231, 256–257
adventure sports. *see also* diving;
        mountaineering 14–15,
        67, 102–103, 119, 135, 167,
        180–181, 187, 209, 219, 231,
        257, 267, 277
air travel 168–169
ancient civilisations & archaeology 67,
        178–179
animal welfare, *see also* conservation
        126, 127, 228
Antarctica 250–251
Argentina 243
    Buenos Aires 39, 131
    Cartagena 63
    Pan-American Highway 270
    Patagonia 197
    Ushuaia 235
art & architecture 22–23, 37, 91, 93
Australia 75, 167, 253
    Byron Bay 205
    Canberra 169
    Christmas Island 113, 265
    Cradle Mountain National Park 129
    Highway 1 271
    Melbourne 93, 149
    New South Wales 225
    Ningaloo Reef 99
    Sunshine Coast 35
    Sydney 15, 37, 125
    Tasmania 66
    Uluru 121, 213, 274
    Victoria 180
    Zaytuna Farm, NSW 273
awe 122–123, 162–163

B

Bahamas
    Andros Islands 29, 123
Belize 143
    Blue Hole 229
    Sarteneja 111

blogging, *see also* journaling 158–159
boats & sailing 94, 132–133, 187, 228,
        253
Bolivia 243, 293
    Salar de Uyuni 123
Botswana 121
Brazil 243
    Blumenau 191
    Pantanal 185
    Ponta Porã 156
    Rio de Janeiro 39, 149
    Salvador 189

C

Cambodia 13
    Choeung Ek Killing Field 87
    Siem Reap 126
camping & eco camps 35, 42–43,
        192–193, 238–239, 272–273
Canada 253
    Athabasca Glacier, Alberta 245
    British Columbia 47
    Chilkoot Trail 33
    Churchill 139
    Fogo Island, Newfoundland 235
    Nova Scotia 59
    Pan-American Highway 270
    Québec 11
    St John's 113
    Whistler Blackcomb 103
    Yukon River 231
Caribbean
    Bonaire 245
    St-Martin & Sint Maarten 249
challenge 10–11, 134–135, 146–147,
        152–153, 156–157, 170–171,
        180–181, 186–187, 202–203,
        206–209, 210–211, 214–215,
        218–219, 224–225, 230–231,
        240–243, 248–249, 252–253,
        254–255, 256–257, 260–261,
        266–267, 268–271, 274–275,
        276–277, 278–279
children (travel with) 68–71
Chile 243
    Patagonia 197

Rapa Nui (Easter Island) 265
    Torres del Paine National Park 55, 193
China
    Chengdu 191
    Chóngqìn 201
    Gobi desert 197
    Guangxi Province 275
    Hong Kong 89, 92, 261
    Huanglong National Park 101
    Khunjerab Pass 156
    Mt Kailash, Tibet 274
    Shànghai 37
    Sichuan's Mt Emei 61
    Silk Road 269
    Taiwan 165
climate change 146–147, 192–193,
        228–229, 236–237, 244–245
Colombia 211, 243
    Cartagena de Indias 109
    Ciudad Perdida 178
    Pan-American Highway 270
    Pijao 29
community 38–39, 60–61, 74–75, 82–
        83, 90–93, 110–111, 126–127,
        190–191, 238–239, 252–253,
        285–286
conservation 127, 142–143, 160–161,
        228–229, 232–233, 244–245
Cook Islands
    Atiu 283
Costa Rica 205
    Lapa Rios 129
creative activities 48–49, 93, 94–95,
        130–131, 158–159
Cuba
    Havana 267
culture 46–47, 82–83, 88–89, 120–121,
        178–179, 260–261
culture shock 172–175
cycling 66, 170–171, 253
Cyprus 167

D

dancing 62–63, 93, 130–131,
        267
dark tourism 86–87

Index

Index

Travel Goals

August 2019
Published by Lonely Planet Global Limited
CRN 554153
www.lonelyplanet.com
10 9 8 7 6 5 4 3 2 1
Printed in China
ISBN 978 1 78868 620 4
US ISBN 978 1 78868 753 9
© Lonely Planet 2019
© photographers as indicated 2019

Managing Director, Publishing Piers Pickard
Associate Publisher Robin Barton
Commissioning Editor Dora Ball
Art Director Daniel Di Paolo
Design & Layout Tina García
Illustrator David Doran
Editors Gabby Inness, Nick Mee
Proofreader Lucy Doncaster
Picture Research Lauren Marchant
Cartography Wayne Murphy
Print Production Nigel Longuet
Thanks Neill Coen, Simon Hoskins, Rebecca Law, Flora Macqueen

Writers Dora Ball, Sarah Barrell, Ray Bartlett, Sarah Baxter, Kerry Christiani, Rucy Cui, Megan Eaves, Janine Eberle, Adam Karlin, Ethan Gelber, Alexander Howard, Anita Isalska, Daniel James Clarke, Patrick Kinsella, Emily Matchar, Omo Osagiede, Matt Phillips, Nora Rawn, Sarah Reid, Simon Richmond, Helena Smith, Clover Stroud, Adam Skolnick, Mara Vorhees, Karla Zimmerman.

STAY IN TOUCH lonelyplanet.com/contact

Australia
The Malt Store, Level 3,
551 Swanston St, Carlton, Victoria 3053
T: 03 8379 8000

USA
124 Linden St, Oakland,
CA 94607
T: 510 250 6400

Ireland
Digital Depot, Roe Lane (off Thomas St),
Digital Hub, Dublin 8 D08 TCV4

Europe
240 Blackfriars Rd,
London SE1 8NW
T: 020 3771 5100